MEMBRANE TRANSPORT PROCESSES
VOLUME 3

MEMBRANE TRANSPORT PROCESSES SERIES

VOLUME 1: Joseph F. Hoffman, editor, 488 pp., 1978.

VOLUME 2: Daniel C. Tosteson, Yu. A. Ovchinnikov, and Ramon Latorre, editors, 472 pp., 1978.

VOLUME 3: Charles F. Stevens and Richard W. Tsien, editors, 168 pp., 1979.

Membrane Transport Processes
Volume 3

Ion Permeation
Through Membrane Channels

Editors

Charles F. Stevens, M.D.
Professor of Physiology
Yale University School of
Medicine
New Haven, Connecticut

Richard W. Tsien, D. Phil.
Associate Professor of
Physiology
Yale University School of
Medicine
New Haven, Connecticut

Raven Press ■ New York

Raven Press, 1140 Avenue of the Americas, New York, New York 10036

Made in the United States of America

Library of Congress Cataloging in Publication Data
Main entry under title:

Ion permeation through membrane channels.

    (Membrane transport processes; v. 3)
    Includes index.
    1. Ion-permeable membranes. 2. Ion exchange.
I. Stevens, Charles F., 1934–   II. Tsien,
Richard W. III. Series.
QH509.M45 vol. 3 [QH601] 574.8'75'08s   77–85069
ISBN 0–89004–236–5   [574.8'75]

Preface

One of the central questions in biology concerns the basis of ion permeation through membranes. Ionic channels are of great interest because they are essential to many activities including excitability, solute and water transport, volume regulation, and sensory transduction. These and many other functions of cellular and subcellular membranes depend on the ability of channels to permit rapid motion of one ion while selecting against other very similar ions. Research over the past few decades has brought us closer to understanding how channels work at the molecular level. Recently it has become clear that many channels are water-filled pores with considerable structural specialization.

This volume presents new research by some of the leading investigators in this field. Several chapters are devoted to studies on various cation channels in nerve and muscle membranes and to characterization of the dimensions and electrical properties of single ionic pores. The work in biological membranes is complemented by research using model systems where the structure of the conducting pathway is known—artificial lipid bilayers containing pore-forming molecules or solid crystals. Two main themes are the importance of subtle features of the conducting structure and the strong possibility of interactions between individual ions. The new developments have provoked important modifications of the theoretical framework for describing ion motion within membranes on the experimental front. Some chapters discuss models based on Eyring rate theory or electrodiffusion and the limitations of these approaches.

This book is aimed at graduate students in the areas of physiology, biophysics, and biochemistry as well as researchers in these areas.

The Editors

Contents

Contributors

David Attwell
University Laboratory of Physiology
Oxford, England OX 1 3PT

Ted Begnisich
Department of Physiology
University of Rochester School of Medicine
and Dentistry
Rochester, New York 14620

Michael Cahalan
Department of Physiology
University of California College of
Medicine
Irvine, California 92717

Vincent Dionne
Department of Physiology and Biophysics
University of Vermont College of Medicine
Burlington, Vermont 05401

Gregory C. Farrington
General Electric Research and Develop-
ment Center
Schenectady, New York 12301

Alan Finkelstein
Departments of Physiology, Neuroscience,
and Biophysics
Albert Einstein College of Medicine
Bronx, New York 10461

D. A. Haydon
Physiological Laboratory
University of Cambridge
Cambridge, England CB2 3EG

Bertil Hille
Department of Physiology and Biophysics
University of Washington School of
Medicine
Seattle, Washington 98195

S. B. Hladky
Physiological Laboratory
University of Cambridge
Cambridge, England CB2 3EG

Peter Läuger
Department of Biology
University of Konstanz
Konstanz, West Germany

Carol A. Lewis
Department of Physiology
Yale University School of Medicine
New Haven, Connecticut 06510

Paul A. Rosenberg
Departments of Physiology, Neuroscience,
and Biophysics
Albert Einstein College of Medicine
Bronx, New York 10461

Charles F. Stevens
Department of Physiology
Yale University School of Medicine
New Haven, Connecticut 06510

Richard W. Tsien
Department of Physiology
Yale University School of Medicine
New Haven, Connecticut 06510

B. W. Urban
Physiological Laboratory
University of Cambridge
Cambridge, England CB2 3EG

Other Participants

Clay Armstrong
University of Pennsylvania

Philippe Ascher
Ecole Normale Superieure, Paris

Ernst Bamberg
University of Konstanz, West Germany

Roland Benz
University of Konstanz, West Germany

Emile L. Boulpaep
Yale University

Josephine Briggs
Yale University

John R. Brockelhurst
Standard Telecommunications Laboratories, Ltd.

Thomas J. Callahan
Yale University

Donald Campbell
Yale University

Vincent Castranova
Yale University

W. Knox Chandler
Yale University

Thomas J. Colatsky
Yale University

Marco Colombini
Albert Einstein College of Medicine

Kenneth A. Doeg
University of Connecticut

Robert S. Eisenberg
Rush Medical College

Bliss Forbush, III
Yale University

Margaret C. Foster
University of California, San Diego

Jeffrey C. Freedman
Yale University

Robert J. French
Biophysics Laboratory, National Institutes of Health

Otto Froehlich
University of Chicago

David C. Gadsby
Rockefeller University

Gerhard Giebisch
Yale University

William Gilly
Yale University

Lawrence Goldman
University of Maryland

Michael E. Green
City College, City University of New York

Robert B. Gunn
University of Chicago

Harold G. Hempling
Medical University of South Carolina

Joseph F. Hoffman
Yale University

Toshihiko Iijima
University of Connecticut Health Center

Meyer Jackson
Yale University

Jack Kaplan
Yale University

Frederic Kavaler
Downstate Medical Center, State University of New York

Peter Kingsley
Cornell University

Thomas Klitzner
University of Pennsylvania

Sally Krasne
University of California, Los Angeles

Edward F. LaBelle
Western Illinois University

Robert LaMotte
Yale University

E. M. Landau
Harvard Medical School

Derek Layton
International Institute of Cellular Pathology, Brussels, Belgium

Robert de Levie
Georgetown University

Eduardo Marban
Yale University

Sandy Markowitz
Yale University

James Maylie
University of Pennsylvania

Stuart McLaughlin
State University of New York at Stony Brook

E. Meymaris
Yale University

Christopher Miller
Yale University

Martin Morad
University of Pennsylvania

Robert J. Myerson
School of Natural Science, Princeton, New Jersey

Erwin Neher
Max-Planck Institute, West Germany

Peter Oosting
Philips Laboratories, Holland

Joaquaim Precopio
Cornell Medical School

Juta K. Reed
University of Toronto

Murdoch Ritchie
Yale University

B. Rudy
Eastern Pennsylvania Psychiatric Institute

Fred Sachs
State University of New York

H. Saito
Yale University

David E. Schafer
Yale University and VA Hospital

Jim Scurlock
Yale University

Jon Shoukimas
Biophysics Laboratory, National Institutes of Health

Brij B. Shrivastav
Duke University Medical Center

Steve Siegelbaum
Yale University

Jane Talvenheimo
Yale University

James B. Wade
Yale University

Jens Warncke
Yale University

Ann Warner
University College, London

Fred S. Wright
Yale University

Rolf Ziskoven
II. Physiological Institute, West Germany

Robert Zwanzig
University of Maryland

Membrane Transport Processes, Volume 3,
edited by C. F. Stevens and R. W. Tsien.
Raven Press, New York, © 1979.

Introduction:
Changing Views of Ion Permeation

Charles F. Stevens and Richard W. Tsien

*Department of Physiology, Yale University School of Medicine,
New Haven, Connecticut 06510*

Our view of how ions move through channels has been changing steadily in the past 10 years, and simultaneously we have been refining our picture of the physical nature of membrane channels. The classic picture of ion permeation, dating from the 1940s, represents a channel as an aqueous pore through which ions can move by diffusion under concentration and voltage gradients. The theoretical approach used for such aqueous pores is the physical chemistry of electrodiffusion developed largely in the last century.

The classic electrodiffusion approach has become increasingly less satisfactory as more refined observations about transport through channels have accumulated. In recent years, mostly since 1972, workers have viewed channels as analogous to one-dimensional crystals and have treated the movement of ions with the same theories used by the solid state physicists for ionic conduction in crystals. This approach simultaneously accounts for a number of phenomena unexplained by electrodiffusion and includes naturally the mechanism of ion selectivity by the channel within a description of ion movement through the channel.

Because our concept of ionic transport through membrane channels has changed so much and because the new concepts are being applied to a variety of different channel types—pores in artificial membranes, and channels in nerve and in the postsynaptic membrane—it seemed to us that a conference bringing together workers who were applying the same concepts to different systems would be beneficial. The conference, sponsored by the Yale Department of Physiology and supported by a grant from the Grass Instrument Company, was held in August, 1977. This book arose from the conference and is intended to survey the new approach to membrane permeation processes, which now seems well established.

The new view represents the membrane channel as a sequence of energy wells and energy barriers; the permeating ions hop—according to a Poission process—from well to adjacent well with a rate that decreases exponentially with the height of the energy barrier over which the ion must hop. Selectivity of the channel arises, in this view, from the different well depths and barrier heights through which ions of the various species must pass. Although the well depths and barrier heights have not yet been calculated from a physical theory, it is clear that they represent electrostatic and other interactions of ions with the channel constituents.

Applying this view to specific situations involves a number of issues. How many wells and barriers are taken to represent the membrane channel? How high are the barrier and wells for the various ionic species? How do ions interact within the channel—that is, may a well be simultaneously occupied by more than one ion and can an ion displace an occupying ion from a well or must jumps occur only into unoccupied wells? What are the physical correlates of the wells and barriers and how can these energies be calculated from information about the channel structure? These questions are approached in the following chapters, which serve to summarize the state of our knowledge on this type of permeation mechanism and to document reasons why workers in the field have abandoned the traditional approach for this newer one.

The book is arranged in four sections, the first dealing with theoretical issues in ion permeation and the remaining three with applications to ionic diffusion in a crystal, the gramicidin channel in artificial bilayers, sodium and potassium channels in nerve membranes, and channels in a postsynaptic membrane opened by neutrotransmitter.

The initial chapter of this book, by Bertil Hille, places the new approach in historical prospective and reviews the experimental evidence—starting from the important Hodgkin and Keynes paper of 1955—that has forced membrane biophysicists away from the traditional macroscopic theories toward the views presented here. Even the simplest realizations of the well–barrier approach for membrane channels lead to very complicated algebraic expressions; for this reason, one is often forced to explore limiting cases. In his chapter, Läuger gives an extensive treatment of one particularly important limiting case—the low ionic concentration limit in which each channel contains at most one ion. For this situation, explicit expressions can be obtained not only for channel conductance and zero-current potential, but also for more complicated qualities such as properties of the fluctuations associated the probabilistic hopping of the ions through the channel. Channels have, in general, nonlinear voltage current relationships that contain information about the underlying permeation mechanisms. Attwell explores, in his chapter, some of the problems associated with extracting this information. For simplicity he adopts a formalism that excludes ionic interactions within the channel, but the effects explored and the general analytic approaches adopted could be extended to more complicated situations.

Ion transport by the hopping mechanism was first proposed for diffusion in crystals, and only in this situation is detailed crystographic evidence about the structure of the medium in which hopping takes place available. Farrington describes studies on ionic conduction in two structural forms of alumina crystals. Such crystals are particularly relevant as models for biological systems because they contain oxygen atoms with spacings similar to those believed to occur in biological membranes, because they have high conductivity for ions, and because movement of biologically relevant species such as the alkali metal ions have been studied. In his contribution, Farrington summarizes information about the structure of beta and beta″ alumina and ionic conduction in these crystals.

The gramicidin molecule can be incorporated into lipid bilayer membranes where it forms ionic channels with dimensions similar to those of biological channels. Because the primary structure of this antibiotic is known and information is available about its conformation in the membrane, many workers have been attracted to studying transport using gramicidin channels as a model system. The hope is that transport properties of this channel can ultimately be understood in terms of physical interactions between the permeating ions and the channel structure. All ionic channels through which ions pass contain water molecules, but little is known about the important interactions between water within the channel, the channel structures, and the permeating ions. Finkelstein and Rosenberg have been studying water and ion fluxes through gramicidin channels and have provided the first picture of water's role within the channel: it appears that water and ions move single file through the gramicidin channel so that one ion and half a dozen water molecules are forced to move as a unit. The gramicidin channel is a particularly favorable one to study because the ionic composition of the bathing medium is readily altered over a large range and because electrical properties of individual channels can be investigated. Studies on the gramicidin channels are replete with examples of phenomena that require ionic interactions within the channel for their explanation, and it is for this channel that hopping models are most fully developed. Hladky, Urban, and Haydon explain why their three-barrier–two-well model for the gramicidin channel is suggested by the structure of the gramicidin molecule and summarize the evidence that such a model does account for the rather complicated properties of gramicidin channels.

The original observation that led to the hopping model was presented in a paper by Hodgkin and Keynes in 1955. They found that the extracellular potassium concentration influenced the outflux of radiolabeled potassium and proposed that potassium ions had to move single file through squid axon potassium channels. According to their analysis, Hodgkin and Keynes found that the potassium channel contained two or three potassium ions that could not slip past one another. Because of some possible technical problems with the original Hodgkin and Keynes experiment, however, it was important to repeat their observation with more modern methods. In his chapter, Begenisich describes a modern version of the Hodgkin and Keynes experiment and shows that the flux interference they reported was not an artifact. The observations reported by Begenisich are consistent with a scheme in which a potassium ion can "knock on"—that is, displace—potassium ions already occupying energy wells. In their chapter on squid axon sodium channels, Begenisich and Cahalan show that here, too, ion interactions within a channel must occur. They examine certain simple schemes involving energy wells and barriers that can accomodate the ionic interaction effects observed in their experiments.

Of the biological channels that have been studied, the acetylcholine-activated channel at the frog neuromuscular junction is probably the best understood and in many ways the simplest. Whereas the axon channels tend to pass sodium ions while excluding potassium ions (or the reverse), opening of channels at the neuromuscular junction causes the simultaneous movement of sodium and potassium. In his chapter, Dionne presents evidence that these ions are subject to identical gating, as though they are both passing through the same channel. Acetylcholine, then, seems to open a single class of channels permeable to both sodium and potassium rather than, as has been proposed, to open two populations of channels, one that passes sodium and one that passes potassium. The selectivity of this channel is relatively slight, with sodium and potassium being nearly equally permeable. Even this relatively unselective channel, however, shows evidence for ion interaction, as described by Lewis and Stevens. A very simple scheme involving a single energy well surrounded by two barriers is sufficient, however, to account for the electrical properties of the acetylcholine-activated channel, whereas the traditional macroscopic approaches are inadequate.

From ionic conduction in crystals through gramicidin channels in artificial membranes to functionally important channels in biological membranes, it seems that the new approach described in this book is required. We hope that this collection of chapters summarizes the current state of knowledge in this field and provides background information to those who wish to extend our information about ionic permeation through channels. We seem to be rapidly approaching the time when the simple description of ionic movements within channels will be relatively accurate so that the next level of analysis, the correlation of structure with energy barrier and well diagrams, can be approached.

Membrane Transport Processes, Volume 3,
edited by C. F. Stevens and R. W. Tsien.
Raven Press, New York, © 1979.

Rate Theory Models for Ion Flow in Ionic Channels of Nerve and Muscle

Bertil Hille

*Department of Physiology and Biophysics, University of Washington School of Medicine,
Seattle, Washington 98195*

The simplest and one of the oldest biophysical representations of the passive permeability of cell membranes to ions is in terms of an ohmic equivalent circuit. The pathways for ionic movements are symbolized by fixed conductors, and the thermal driving forces on ions are symbolized by an electromotive force in series with the conductor. This equivalent circuit has been extremely useful in studying the cable properties of extended cells and in the analyses of electrical responses of excitable cells. However, in our present concern with the mechanisms of permeation at the molecular level, the electric circuit analog does not answer the question of how the ions are actually moving. We need instead a theoretical framework that can describe ionic motions in terms of the structure of the permeable pathways across the membrane.

Excitable membranes contain several types of ionic channels, each of which is a macromolecular pore permeable enough to pass several millions of ions in one second and yet narrow enough to exhibit at least a modest selectivity in the charge and size of the ions allowed to pass. These ionic fluxes have been modeled after two major types of passive transport theories: free diffusion (16,29), where the ions move continuously through a relatively structureless medium, and rate theory (12), where the ions hop stepwise across a series of discrete energy barriers. The purpose of this chapter is to show that many membrane phenomena are incompatible with free diffusion in ionic channels and that certain versions of the rate-theory approach have the potential to describe most of what is known. For relatively narrow ionic channels, rate theory may give us enough structural insights to set up a realistic molecular dynamic simulation of ionic permeation. In any case, ion flows in ionic channels are correctly termed "diffusion" or "electrodiffusion" even though they are constrained by more complicated particle correlations and interactions with the channel than are implied by the original concept of free diffusion.

FREE DIFFUSION

In the standard version of free diffusion for ionic permeation (1,16,29,33), everything except possibly the moving ion is represented as a continuum with time-averaged properties and the movements of each ion species are usually described by the Nernst-Planck equations. Thus, the driving forces are expressed as the gradients of the electrochemical potential. The walls of the diffusion path may be sharply defined, and the standard chemical potential of the diffusing ion might even vary in complicated ways with position in the channel, but the solvent and all other ions feature only through their time-averaged effects on electrochemical potential and mobility, and not through specific instantaneous effects as discrete particles. These theories are fundamentally continuum models. In the absence of any bulk flow of solvent, such free-diffusion systems always obey the Ussing (41) flux-ratio criterion. Namely, the ratio $\vec{J}_S/\overleftarrow{J}_S$ of unidirectional fluxes (measured by tracers) of an ion, S, is equal to the ratio of electrochemical activities of that ion:

$$\vec{J}_S/\overleftarrow{J}_S = \frac{[S]_i}{[S]_o} \exp\ (zEF/RT) \qquad [1]$$

where $[S]_i$ is inside and $[S]_o$ outside ionic activity, E is the membrane potential, z the valence of the ion, R the gas constant, T absolute temperature, and F, the Faraday. This result follows whenever an ion coming from the right experiences the same position-dependent forces in the membrane as an ion coming from the left. Some free-diffusion systems also satisfy independence (30), in which the chance that an ion moves across the membrane is independent of all other ions. This condition requires that changing the concentration of bathing ions not change the forces on an ion in the membrane. For example, neither the surface potentials nor the space charge in the membrane may be affected. These conditions might be obtained with thin (<10 nm) neutral membranes at low ion concentrations.

Finally, for thin membranes that are also homogenous, the electric field in the membrane would be constant from one side to the other, and fluxes of ions would obey the Goldman (16) and Hodgkin and Katz (29) current equation, also called the constant-field flux equation. For example, the current I_{Na} of Na^+ ions would be

$$I_{Na} = E\frac{F^2}{RT}\,P_{Na}\,\frac{[Na]_o - [Na]_i \exp(FE/RT)}{1 - \exp(FE/RT)} \qquad [2]$$

where P_{Na} is a voltage-independent and concentration-independent permeability coefficient. When two cations share the same pathway, then their permeability ratio P_{Na}/P_K can be determined from the zero-current potential or reversal potential, E_r, when the membrane is bathed by both ions (16,29).

$$E_r = \frac{RT}{F} ln\left(\frac{P_{Na}[Na]_o + P_K[K]_o}{P_{Na}[Na]_i + P_K[K]_i}\right) \qquad [3]$$

This widely used equation may be derived directly, using Eq. [2] for the homogenous, constant-field membrane, or, alternatively, from other much less restrictive conditions (23). In addition, Eq. [3] is often used even in conditions where the *P*s are not constants, simply as an empirical definition of permeability ratios for a given experimental condition. Equations [2] and [3] have been the mainstays of bioelectrochemistry for 30 years.

STUDIES ON IONIC CHANNELS

Although they have played an absolutely essential role in the first tests of the ionic hypothesis, in subsequent work free-diffusion theories have been found inadequate for describing a variety of properties of ionic fluxes in biological channels. Consider first the delayed rectifier K channel of axons and the inward rectifier K channel of muscle. Hodgkin and Keynes (31) found, as early as 1955, that the Ussing flux-ratio criterion is not obeyed by K^+ fluxes in axons. Instead the fluxes satisfy the equation

$$\frac{\vec{J}_K}{\overleftarrow{J}_K} = \left\{ \frac{[S]_i}{[S]_o} \exp(EF/RT) \right\}^{n'} \qquad [4]$$

with $n' = 2.5$. Horowicz et al. (32) obtained a similar result with K^+ fluxes in skeletal muscle, with $n' = 2$. Systems fitting Eq. [4] with $n' > 1$ have positive coupling between movements of individual ions, almost as if the ions move in multimolecular packets of an average size n'. Such coupling violates the basic tenets of free diffusion and cannot be described by that type of theory. Any appropriate theory must invoke the simultaneous presence of two or more ions in a pore at a time, which move in a correlated fashion.

In many cases the properties of an ionic channel are like those of a chain of ionic binding sites capable of saturation rather than like those of a system where ions move independently in a continuum. Evidence for ionic binding sites in K channels comes from the effect of changing the intracellular or extracellular bathing cations on outward K^+ current or outward unidirectional $^{42}K^+$ fluxes. Replacing some of the intracellular K^+ ions in a squid axon by Li^+, Na^+, Cs^+, or a variety of other small cations reduces outward K^+ current by much more than is expected from the small reduction of $[K]_i$ and in a manner that is steeply voltage dependent (4,6,14). Similar results have been obtained with the node of Ranvier (3,13,23). Two examples are given in Fig. 1, which shows the current-voltage relations for K^+ movement across a node of Ranvier. The K^+ currents are recorded while Na^+ ions or Cs^+ ions are gradually appearing at the nodal axoplasm by diffusion from a cut in a neighboring internode. Within minutes, the outward currents become strongly depressed, showing that Na^+ and Cs^+ ions are specific blocking agents that block with a steep voltage dependence. The common interpretation is that the "foreign" cations may be driven partway through the channel from the inner end and become "stuck," at some strong binding site or at some particularly high barrier within the channel. With large depolarizations, the stuck ion may actually be pushed right out to the external medium giving a relief of block at very positive potentials (14).

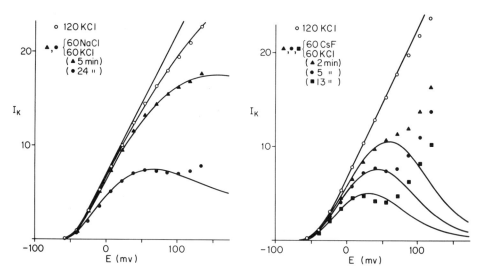

FIG. 1. Voltage-dependent block of K channels in myelinated nerve by Na^+ and Cs^+ ions. The points are the steady-state outward current in K channels of a node of Ranvier stepped to different potentials under voltage clamp. Initially *(open circles)* the ends of the fiber were cut in 120 mM KCl, but then this solution was replaced by 60 mM NaCl + 60 mM KCl or 60 mM CsF + 60 mM KCl. As the Na^+ or Cs^+ ions diffuse into the nodal axoplasm, K channels become blocked. Smooth curves are derived using the single-site, blocking model of Woodhull (43), assuming a blocking site 80% across the membrane field and that Na^+ ions are slightly permeant and Cs^+ ions impermeant. The model does not take into account ionic interactions in the channel and does not give a good fit to the observations. Experiments from 6/24/72 and 7/4/72 are done in the same series as those of refs. 22 and 24.

In a different kind of experiment, adding external K^+ decreases the efflux of $^{42}K^+$ from squid giant axons (2), whereas the same maneuver can *increase* the $^{42}K^+$ efflux from frog skeletal muscle (28,32). Hence, in one case, external K^+ ions block the outward movement of K^+ ions, and in the other, they may relieve a block. These various deviations from independent movement of ions can readily be discussed in terms of competitive binding of K^+ ions and other ions to discrete sites within the K channel, but cannot be described in terms of free diffusion.

Similar but less pronounced deviations from independence and blocking effects of external foreign cations have been studied with Na channels as well. For example, when the concentration of external Na^+ ions is raised from 58 to 232 mM, around the node of Ranvier, the peak inward Na^+ current increases, but only to 70% of the size expected from independence, and when permeant Tl^+ or guanidinium ions are added to a solution containing 58 mM Na^+, the net inward current actually decreases (23,24) although the reversal potential increases. Protons also block Na^+ currents as if there is an ionized acid group with a pK_a of 5.2 to 5.4 in the channel that is essential for permeation (21,43).

Yet another problem for free-diffusion theories of Na and K channels is that measured permeability ratios, defined by Eq. [3] for reversal potentials, vary depending on the ionic content of the solutions. In Na channels of the squid, P_{Na}/P_K can be varied from 13 to 4 by lowering the internal K^+ concentration (5,6), and in inward rectifier K channels of echinoderm eggs, P_K/P_{Tl} can be varied from >5 to 0.8 by increasing the proportion of Tl^+ in a K^+–Tl^+ mixture bathing the egg (17). These changes of permeability could be accounted for by a hybrid model in which ions binding to the surfaces of the membrane cause structural changes of a free-diffusion path through the membrane, but they are even more easily accounted for by a rate-theory model that requires competition for sites but no structural changes (25).

RATE-THEORY MODELS

Danielli (8,9) was one of the first to describe membrane permeation in terms of stepwise crossing of a series of energy barriers, as in Fig. 2. These ideas received additional precision from developments in rate theory.

Chemical rate theory accounts for the rate constants of elementary steps in chemical reactions in terms of the potential energy changes of the reactants as they approach one another and undergo the transformation being considered (15). In its simplest form and in analogy to the earlier Arrhenius theory, the fundamental concept is that the reactants enter into a high-potential-energy "activated complex" or "transition state" that then rapidly breaks down to the products (or to the initial reactants) and that the rate constant, k, is the negative exponential of the free energy ΔG of this activated complex relative to the starting point

$$k = Q \exp(-\Delta G^{\ddagger}/RT) \qquad [5]$$

where Q is a frequency factor, related to the frequency of atomic vibrations, and $\Delta G^{\ddagger}$ is called the activation free energy.

Eyring and his colleagues formalized many of these concepts and proceeded

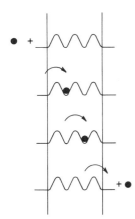

FIG. 2. A particle crossing the membrane by jumping over a sequence of energy barriers. The fundamental difference between rate-theory models and continuum models is that in rate theory the forward motion is a discontinuous jump over a finite distance rather than a continuous migration down a gradient.

to apply them to a wide variety of rate processes including bioluminescence, protein denaturation, narcosis, and diffusion (34). By 1949 they used this approach to derive an equation for the current flowing across nerve membranes (12), which, for a large number of similar elementary steps, becomes identical to the Goldman-Hodgkin-Katz equation, Eq. [2] (42). Like Danielli (8,9), Eyring et al. (12,42) viewed the membrane as a multicompartment system, as in Fig. 3A, where the C's denote the concentration of the diffusing particle in the bathing compartments and the internal "compartments." Rate theory then provides the rate constants for transfer between the compartments from the heights of the energy barriers between them. The new feature added by Eyring et al. (12) was the inclusion of the effect of the membrane field on the energy barrier (activation energy) by adding the electrostatic potential of an ion in the field to the field-independent energy barriers. In this way, the rate constants become voltage dependent with the form

$$k_i = b_i \cdot \exp(-\alpha_i zEF/RT) \tag{6}$$

where b_i is the rate constant at $E = O$ mV and α_i is a positive or negative number equal to the fraction of the potential difference E crossed by the ion in reaching the transition state.

In subsequent applications of rate theory to biological membrane transport, the important features have been the representation of diffusion as progress through a multibarrier system and that the voltage dependence comes into the rate constants as an exponential term. The energy profiles have served more

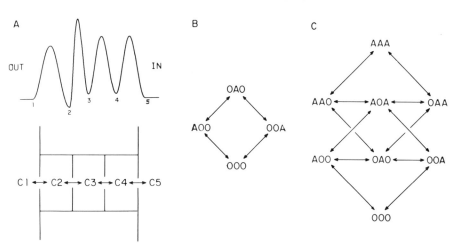

FIG. 3. A four-barrier energy diagram of diffusion and the state diagrams of three different types of kinetic models corresponding to it. The arrows represent permitted transitions among states. **A:** Linear, multicompartment kinetic model where Cs represent concentrations of diffusing substance. There is no implied limit on how high the concentrations can be. **B:** One-ion, three-site channel in presence of permeant ion species A. Each site can be either empty "O" or occupied "A," but only one site may be occupied at a time. **C:** Multi-ion, three-site channel. Total occupancy of the channel may range from zero to three. (After Hille, ref. 23.)

of a pictorial purpose than a fundamental one, so far, since neither can we determined them very well from experiments nor do we know how they would be related to the chemical structure of the pore. Thus from a practical point of view the absolute-reaction-rate part of the theory has not been applied, but the philosophy has been extremely helpful.

The multicompartment kinetics (Fig. 3A) used by Eyring et al. (12,42) for membrane diffusion correspond to a linear system of equations and in this sense are equivalent to free diffusion, which returns to the many problems already discussed. For example, the fluxes derived by these multicompartment kinetics obey the Ussing flux ratio and independence, and the permeability ratios for different ions do not depend on ionic compositions. To go further, it is necessary to abandon the idea of time-averaged compartments and to think instead of a pore with discrete, internal, saturable sites where transitions among different states of occupancy constitute transport. This idea was clearly expressed by Hodgkin and Keynes (31) and brought into the currently used form by the important work of Heckmann (18–20). As an elementary example, consider an empty three-site channel, symbolized 000, which picks up an ion, A, from the right side and discharges it on the left by the following sequence of transitions: 000 + A → 00A → 0A0 → A00 → A + 000, each representing a single crossing of an energy barrier and each described by rate theory. These transitions are conveniently summarized as a reaction diagram or kinetic state diagram in Fig. 3B, which, because of the explicit absence of multiply occupied states like A0A or AAA, represents what can be called a one-ion pore. Formally the kinetics in such one-ion pores are identical to simple enzyme kinetics with transitions of enzyme E, substrate S, and products P as follows E + S → ES → ES' → EP → E → P, and the resulting kinetic behavior is, therefore, well known. With one substrate, the fluxes saturate at high concentrations, and with a second substrate there is competitive inhibition of the flux of the first by the second. Thus, as Läuger has shown in detail, one-ion pore models, which show a conventional Michaelis-Menten saturation behavior, are quite easy to develop mathematically (22,23,35,38).

The one-ion kinetic model of Fig. 3B has been applied successfully to Na channels to describe the deviations from independence on addition of extra external Na^+ ions or foreign cations (7,24,38). However, it is still too simple to account for concentration-dependent permeability ratios or for deviations from the Ussing flux ratio, described by Eq. [4] (35,38). The second of these properties requires the simultaneous presence of more than one ion at a time in the channel, and the first is also conveniently explained this way. Hodgkin and Keynes (31), Heckmann (18–20), and many subsequent investigators have examined models in which the list of allowed states of the pore includes those with multiple occupancy and in which ions in the channel cannot pass each other, conditions commonly referred to as giving "single-file diffusion." Multiple occupancy is reasonable since, if there are several binding sites in the channel, it should be just a matter of raising the concentration of bathing ions sufficiently to have more than one site occupied. And the no-passing condition is reasonable, as we believe that highly ion-selective channels are relatively narrow pores, and, therefore, mutually repelling ions are not likely to be able to pass each other. Figure 3C gives an example of the kinetic state diagram for a three-site channel bathed by a single type of permeant ion. There are eight states and 24 rate constants (not all independent).

The properties of single-file models can be exceedingly complex and have certainly not been entirely elucidated. Fortunately, the simplest such model with two sites and mirror symmetry has been worked out in analytical detail (18,19,25,26,38) and can be used to illustrate several general properties. Figure 4 shows occupancy and conductance calculations for a hypothetical two-site pore with a low central barrier (see energy profile drawn in inset) and bathed by symmetrical solutions of K^+ ions. As the activity of K^+ is increased, empty channels are converted into singly occupied channels, and then singly occupied ones are converted into doubly occupied ones. The loading of the channel is spread over a very broad activity range in this example because an electrostatic repulsion has been postulated that slows the entrance of a second ion 10-fold and speeds its exit 10-fold as well. The average conductance varies in a nonmonotonic manner as the channels are gradually converted from an all-empty to an all-full condition. At very low activities, the average conductance rises linearly

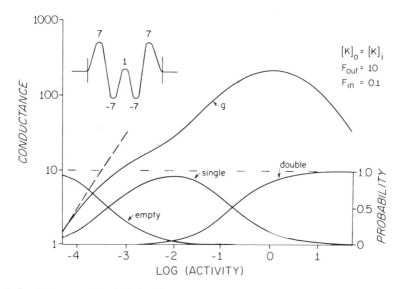

FIG. 4. Conductance and probability of occupancy states of a two-site, single-file pore bathed with equal concentrations of K^+ ions inside and out. The energy profile seen by the first ion entering is represented in the inset, with energies given in multiples of *RT*. A second ion enters with a rate constant 0.1 times that of the first, and then either ion leaves with a rate constant 10 times that for a single ion. The K^+ activity of the symmetrical solutions is increased over a range of six orders of magnitude. (Calculations done by the methods of Hille and Schwarz, ref. 25.)

(dashed line) with activity, but then the rate of increase slows when most channels become singly occupied and few empty channels remain. However, at higher activities, doubly occupied channels begin to be formed, and the conductance begins to rise more steeply once again because the exit rate from these channels is 10-fold faster than from singly occupied ones. Finally, as most channels become full, the conductance actually decreases because vacancies do not last long enough for an ion to cross the central barrier and the only ion motions occurring are rapid but fruitless exchanges of ions between each lateral site and its neighboring bath. This final decrease of conductance is beginning to be referred to as "self block" and arises in a vacancy-diffusion mechanism whenever there are no vacancies left. The same calculations for the same model are shown again in Fig. 5, but now with conductance on a linear scale and with average occupancy rather than the distribution among states. In addition the figure shows the calculated flux-ratio exponent, n', which rises from 1.0 to a peak of 1.91 and returns to 1.0. Hence these models show the flux-coupling phenomenon first discovered by Hodgkin and Keynes (31) in giant axons and first attributed by them to single-file diffusion.

Wolfgang Schwarz and I (25) have reviewed other properties of multi-ion channels and discussed their relevance to biological potassium channels in particular. We give calculations showing concentration-dependent permeability ratios, so-called anomalous mole-fraction dependence of conductances and permeabilities, and steeply voltage-dependent blocks or "rectification" caused by impermeant blocking ions in the internal or external medium. The blocks may be reversed by adding permeant ions to the side opposite from the blocking ion and can be made to imitate closely the properties of inward-going rectification

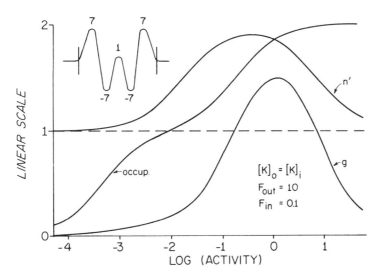

FIG. 5. Conductance, total occupancy, and flux-ratio exponent for the same pore model as in Fig. 4. (Flux-ratio exponent calculated from Eq. [12] of Hille and Schwarz, ref. 25.)

in muscle and egg membranes. We did not attempt to model more complicated phenomena like the block with internal Cs^+ or Na^+ that reverses at high potentials as in Fig. 1. French and Wells (14) describe this phenomenon in detail and show that Na^+ ions can actually pass through potassium channels at high potentials, although they seem impermeant at lower potentials.

It is likely that the first really well-characterized, multi-ion channel will be the gramicidin A channel rather than one of the channels of excitable cells. This model system, brought into prominence by Hladky and Haydon's (1972) discovery of unitary conductance events, shows concentration-dependent permeability ratios, anomalous mole-fraction dependence of conductance in $Tl^+ - Na^+$ mixtures, self-block with Cs^+, and a clear two-step rise of conductance as the bathing concentration of Tl^+ and other ions is increased (10,11,26,27,39,40). Several two-site and four-site rate-theory models have been discussed to explain these phenomena (10,11,26,37). The important advantages of this system over physiological ones now available are that one can measure the properties of a single pore and that the conditions of the system may be varied in the extreme. The chemical structure of the molecule can be varied, the lipid can be varied, the pH can be changed by at least 8 units, and ion concentrations can be studied over six orders of magnitude. These important studies and the theoretical description they produce have and will give us new ideas to apply to the study of ionic channels in excitable cells.

THE FUTURE

To date there have been no serious attempts to make a quantitative fit of a single-file model to the actual transport properties of a biological channel, although in simpler systems, such as the gramicidin channel, much progress has been made. However, as the available measurements become more comprehensive and as our methods and understanding of such models improve, we can soon expect significant advances in understanding of fluxes in physiological channels. I feel that most phenomena now known should be describable by the rate-theory method. Yet, new serious deficiencies of the new models will ultimately be found. These may arise because ionic motion through thermally agitated solvent molecules in a narrow three-dimensional volume defined by a thermally agitated channel macromolecule cannot be mapped perfectly onto a scheme with a one-dimensional, fixed-barrier profile. With the insight gained from barrier modelling and possible direct structural information, the final step will probably be to solve the problem explicitly with solvent molecules, solute particles, and channel in the very powerful, but slow, molecular dynamics method (36).

ACKNOWLEDGMENT

It is a pleasure to acknowledge discussions with Clay M. Armstrong and collaboration with Wolfgang Schwarz that have clarified some questions in this chapter.

This work is supported by grants NS 08174 and FR 00374 from the U.S. Public Health Service.

REFERENCES

1. Adrian, R. H. (1969): Rectification in muscle membrane. *Prog. Biophys. Mol. Biol.,* 19:339–369.
2. Begenesich, T., and DeWeer, P. (1977): Ionic interactions in the potassium channel of squid giant axons. *Nature,* 269:710–711.
3. Bergman, C. (1970): Increase of sodium concentration near the inner surface of the nodal membrane. *Pfluegers Arch.,* 217:287–302.
4. Bezanilla, F., and Armstrong, C. M. (1972): Negative conductance caused by entry of sodium and cesium ions into the potassium channels of squid axons. *J. Gen. Physiol.,* 60:588–608.
5. Cahalan, M., and Begenisich, T. (1976): Sodium channel selectivity. Dependence on internal permeant ion concentration. *J. Gen. Physiol.,* 68:111–125.
6. Chandler, W. K., and Meves, H. (1965): Voltage clamp experiments on internally perfused giant axons. *J. Physiol. (Lond.),* 180:788–820.
7. Chizmadjev, Yu. A., and Aityan, S. Kh. (1977): Ion transport across sodium channels in biological membranes. *J. Theor. Biol.,* 64:429–453.
8. Danielli, J. F. (1939): The site of resistance to diffusion through the cell membrane, and the role of partition coefficients. *J. Physiol. (Lond.),* 96:2P.
9. Davson, H., and Danielli, J. F. (1943): *The Permeability of Natural Membranes,* pp. 310–352. Cambridge Univ. Press, Cambridge.
10. Eisenman, G., Sandblom, J. P., and Neher, E. (1977): Ionic selectivity, saturation, binding, and block in the gramicidin A channel: a preliminary report. In: *Metal-Ligand Interactions in Organic Chemistry and Biochemistry,* Part 2, edited by B. Pullman and N. Goldblum, pp. 1–36. D. Reidel, Dordecht-Holland.
11. Eisenman, G., Sandblom, J. P., and Neher, E. (1978): Interactions in cation permeation through the gramicidin channel: Cs, Rb, K, Na, Li, Tl, H, and effects of anion binding. *Biophys. J.,* 22:307–340.
12. Eyring, H., Lumry, R., and Woodbury, J. W. (1949): Some applications of modern rate theory to physiological systems. *Record Chem. Prog.,* 10:100–114.
13. Frankenhaeuser, B., and Århem, P. (1975): Steady state current rectification in potential clamped nodes of Ranvier. *(Xenopus laevis). Philos. Trans. R. Soc. Lond. [Biol.],* 270:515–525.
14. French, R. J., and Wells, J. B. (1977): Sodium ions as blocking agents and charge carriers in the potassium channel of the squid giant axon. *J. Gen. Physiol.,* 70:707–724.
15. Glasstone, S., Laidler, K. J., and Eyring, H., (1941): The theory of rate processes. McGraw-Hill, New York.
16. Goldman, D. (1943): Potential, impedance, and rectification in membranes. *J. Gen. Physiol.,* 27:37–60.
17. Hagiwara, S., Miyazaki, S., Krasne, S., and Ciani, S. (1977): Anomalous permeabilities of the egg cell membrane of a starfish in $K^+ - Tl^+$ mixtures. *J. Gen. Physiol.,* 70:269–281.
18. Heckmann, K. (1965): Zur Theorie der "single file"—Diffusion, I. *Z. Phys. Chem.* [N. F.], 44:184–203.
19. Heckmann, K. (1965): Zur Theorie der "single file"—Diffusion, II. *Z. Phys. Chem.* [N. F.], 46:1–25.
20. Heckmann, K. (1972): Single file diffusion. *Biomembranes,* 3:127–153.
21. Hille, B. (1968): Charges and potentials at the nerve surface: Divalent ions and pH. *J. Gen. Physiol.,* 51:221–236.
22. Hille, B. (1973): Potassium channels in myelinated nerve: Selective permeability to small cations. *J. Gen. Physiol.,* 61:669–686.
23. Hille, B. (1975): Ionic selectivity of Na and K channels of nerve membranes. In: *Membranes: A Series of Advances,* Vol. 3, edited by G. Eisenman, pp. 255–323. Dekker, New York.
24. Hille, B. (1975): Ionic selectivity, saturation and block in sodium channels. A four barrier model. *J. Gen. Physiol.,* 66:535–560.
25. Hille, B., and Schwarz, W. (1978): Potassium channels as single-file pores. *J. Gen. Physiol.,* 72 *(in press).*

26. Hladky, S. B. (1972): The Mechanism of Ion Conduction in Thin Lipid Membranes Containing Gramicidin A. Ph.D. Thesis. University of Cambridge, Cambridge, England.
27. Hladky, S. B., and Haydon, D. A. (1972): Ion transfer across lipid membranes in the presence of gramicidin A. I. Studies of the unit conductance channel. *Biochim. Biophys. Acta,* 274:294–312.
28. Hodgkin, A. L., and Horowicz, P. (1959): The influence of potassium and chloride ions on the membrane potentials of single muscle fibers. *J. Physiol. (Lond.),* 148:127–160.
29. Hodgkin, A. L., and Katz, B. (1949): The effect of sodium ions on the electrical activity of the giant axon of the squid. *J. Physiol. (Lond.),* 108:37–77.
30. Hodgkin, A. L., and Huxley, A. F. (1952): Currents carried by sodium and potassium ions through the membrane of the giant axon of *Loligo. J. Physiol. (Lond.),* 116:449–472.
31. Hodgkin, A. L., and Keynes, R. D. (1955): The potassium permeability of a giant nerve fibre. *J. Physiol. (Lond.),* 128:61–88.
32. Horowicz, P., Gage, P. W., and Eisenberg, R. S. (1968): The role of the electrochemical gradient in determining potassium fluxes in frog striated muscle. *J. Gen. Physiol.,* 51:193s–203s.
33. Jack, J. J. B., Noble, D., and Tsien, R. W. (1975): *Electric Current Flow in Excitable Cells,* pp. 225–260. Clarendon Press, Oxford.
34. Johnson, F. H., Eyring, H., and Polissar, M. J. (1954): *The Kinetic Basis of Molecular Biology.* Wiley, New York.
35. Läuger, P. (1973): Ion transport through pores: a rate theory analysis. *Biochim. Biophys. Acta,* 311:423–441.
36. Levitt, D. G. (1973): Kinetics of diffusion and convection in 3.2 Å pores. Exact solution by computer simulation. *Biophys. J.,* 13:186–206.
37. Levitt, D. G. (1978): Electrostatic calculations for an ion channel. II. Kinetic behavior of the gramicidin A channel. *Biophys. J.,* 22:221–248.
38. Markin, V. S., and Chizmadjev, Y. A. (1974): *Induced Ionic Transport.* Nauka, Moscow. *(Russ.).*
39. Neher, E. (1975): Ionic specificity of the gramicidin channel and the thallous ion. *Biochim. Biophys. Acta,* 401:540–544.
40. Neher, E., Sandblom, J., and Eisenman, G. (1978): Ionic selectivity, saturation, and block in gramicidin A channels. II. Saturation behavior of single channel conductances and evidence for the existence of multiple binding sites in the channel. *J. Membr. Biol.,* 40:97–116.
41. Ussing, H. H. (1949): The distinction by means of tracers between active transport and diffusion. *Acta Physiol. Scand.,* 19:43–56.
42. Woodbury, J. W. (1971): Eyring rate theory model of the current-voltage relationships of ion channels in excitable membranes. In: *Chemical Dynamics: Papers in Honor of Henry Eyring,* edited by J. O. Hirschfelder, pp. 601–617. Wiley, New York.
43. Woodhull, A. M. (1973): Ionic blockage of sodium channels. *J. Gen. Physiol.,* 66:669–686.

Membrane Transport Processes, Volume 3,
edited by C. F. Stevens and R. W. Tsien.
Raven Press, New York, © 1979.

Transport of Noninteracting Ions Through Channels

P. Läuger

Department of Biology, University of Konstanz, Konstanz, West Germany

BARRIER MODEL OF THE CHANNEL

An ion channel is a structural element that offers to an ion an energetically favorable pathway through the apolar core of a lipid membrane. Such a channel may be represented by a sequence of "binding sites" that are separated by activation energy barriers (Fig. 1). The binding sites are the minima in the potential energy profile that result from interactions of the ion with one or several ligand groups of the channel. A simple situation arises when the ion concentration is low enough so that the probability of finding more than one ion at a time in the channel is vanishingly small (20). This case in which ion–ion interactions within the channel may be neglected is treated in the following.

We assume that the membrane contains N_c channels and that the channels are permeable to a single ion species present in the external phases in concentrations c' and c'' (Fig. 1). If N_o is the average number of ions (referred to total membrane area) located in the outermost energy minimum at the lefthand mouth of the channel, then $p_o = N_o/N_c$ is the probability that, for a given channel, the outer minimum is occupied by an ion. This probability is assumed to be voltage independent and proportional to the aqueous ion concentration; a similar statement applies to the righthand mouth of the channel. Thus,

$$p_o = vc' \; ; \; p_{n+1} = vc'' \qquad [1]$$

where v is a proportionality constant. We denote the probability of finding the i-th minimum in a given channel occupied by $p_i(p_i \ll 1)$. The average net ion flux Φ_i in the single channel over the i-th barrier is then given by (33):

$$\Phi_i = k^l_{i-1} p_{i-1} - k^{ll}_i p_i \qquad [2]$$
$$(i = 1, 2, \ldots, n + 1)$$

where k^l_{i-1} and k^{ll}_i are the voltage-dependent rate constants for jumps over the i-th barrier from left to right and from right to left (compare Fig. 1). In the equilibrium state ($\Phi_i \equiv 0$), the unidirectional fluxes over the i-th barrier in either direction are equal:

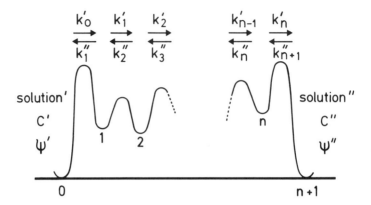

FIG. 1. Potential energy of an ion in the channel. k_i' and k_i'' are the rate constants for jumps from the i-th energy minimum to the right and to the left, respectively. c' and c'' are the concentrations of the permeable ion species, and ψ' and ψ'' are the electrical potentials in the two aqueous solutions.

$$\bar{k}_{i-1}'\,\bar{p}_{i-1} = \bar{k}_i''\,\bar{p}_i \equiv F_i \qquad [3]$$

where $\bar{k}_i'$, $\bar{k}_i''$, and $\bar{p}_i$ are the equilibrium values of k_i', k_i'', and p_i, respectively, and F_i is the unidirectional flux over the i-th barrier at equilibrium. Relation [3] represents a system of $(n+1)$ equations that is easily solved, starting with either $i = 1$ or $i = n+1$, to give

$$\bar{p}_i = p_o \frac{\bar{k}_o'\,\bar{k}_1' \;\ldots\; \bar{k}_{i-1}'}{\bar{k}_1''\,\bar{k}_2'' \;\ldots\; \bar{k}_i''} = p_{n+1} \frac{\bar{k}_{i+1}''\,\bar{k}_{i+2}'' \;\ldots\; \bar{k}_{n+1}''}{\bar{k}_i'\,\bar{k}_{i+1}' \;\ldots\; \bar{k}_n'} \qquad [4]$$

(As the probabilities p_o and p_{n+1} that the outer minima are occupied are assumed to be constant, we may write $\bar{p}_o = p_o$ and $\bar{p}_{n+1} = p_{n+1}$). Equation [4] yields, in accordance with the principle of microscopic reversibility, the relationship

$$p_o\bar{k}_o'\,\bar{k}_1' \;\ldots\; \bar{k}_n' = p_{n+1}\bar{k}_1''\,\bar{k}_2'' \;\ldots\; \bar{k}_{n+1}'' \equiv H \qquad [5]$$

The equilibrium voltage V_e of the permeable ion (valency z) is given by

$$\frac{c''}{c'} = \frac{p_{n+1}}{p_o} = \exp(zu_e) \qquad [6]$$

$$u_e = \frac{V_e}{kT/e_o} = \frac{(\psi' - \psi'')_e}{kT/e_o} \qquad [7]$$

where k is the Boltzmann constant, T, the absolute temperature, e_o, the elementary charge, and ψ' and ψ'', the electrical potentials in the lefthand and righthand aqueous phase (Fig. 1). Equations [5] and [6] together give

$$\frac{\bar{k}_o'\,\bar{k}_1' \;\ldots\; \bar{k}_n'}{\bar{k}_1''\,\bar{k}_2'' \;\ldots\; \bar{k}_{n+1}''} = \exp(zu_e) \qquad [8]$$

As any value of the equilibrium voltage u_e can be achieved by a suitable choice of c' and c'' and as the rate constants k_i' and k_i'' are independent of c' and c'', a relation of the form of Eq. [8] holds at arbitrary voltages $u = (\psi' - \psi'')kT/e_o$:

$$\frac{k_o' k_1' \ldots k_n'}{k_1'' k_2'' \ldots k_{n+1}''} = \exp(zu) \tag{9}$$

ION PERMEABILITY AND ELECTRICAL CONDUCTANCE OF THE CHANNEL

We consider the stationary state of the channel where the ion fluxes over all barriers are equal ($\Phi_i \equiv \Phi$). Relation [2] then represents a system of ($n + 1$) equations for the ($n + 1$) unknown quantities $\phi, p_1, p_2, \ldots, p_n$. The solution for Φ reads (after introduction of Eq. [9]) (20,22):

$$\Phi = P \frac{zu/2}{\sinh(zu/2)} [c' \exp(zu/2) - c'' \exp(-zu/2)]$$

$$= Pzu \frac{c' \exp(zu) - c''}{\exp(zu) - 1} \tag{10}$$

Equation [10] has the form of Goldman's flux equation (11,17). P is the voltage-dependent permeability coefficient of the single channel given by:

$$P = \frac{1 - \exp(-zu)}{zu} \frac{vk_o'}{1 + \sum\limits_{v=1}^{n} s_v} \tag{11}$$

$$S_v = \frac{k_1'' k_2'' \ldots k_v''}{k_1' k_2' \ldots k_v'} \tag{12}$$

At zero voltage, Eq. [10] and [11] reduce to

$$\Phi = P(c' - c'') \tag{13}$$

$$P = \frac{v \tilde{k}_o'}{1 + \sum\limits_{v=1}^{n} \tilde{S}_v} \equiv P_o \tag{14}$$

where $\tilde{k}_o'$ and $\tilde{S}_v$ are the values of k_o' and S_v for $u = 0$.

As the ions move independently through the channel, the unidirectional fluxes Φ' and Φ'' are simply obtained by applying Eq. [10] to the case $c'' = 0$ or $c' = 0$:

$$\Phi = \Phi' - \Phi'' \tag{15}$$

$$\Phi' = Pc' \frac{zu}{1 - \exp(-zu)} \qquad [16]$$

$$\Phi'' = Pc'' \frac{zu}{\exp(zu) - 1} \qquad [17]$$

Equations [16] and [17] may be used to determine the permeability coefficient P from tracer flux experiments.

According to the theory of absolute reaction rates (10,28,32,33), the voltage-dependence of the rate constants may be expressed by

$$k_i' = \bar{k}_i' \exp(\alpha_{i+1} zu/2) \qquad [18]$$
$$k_i'' = \bar{k}_i'' \exp(-\alpha_i zu/2) \qquad [19]$$

$\bar{k}_i'$ and $\bar{k}_i''$ are the values of k_i' and k_i'' at zero voltage and α_i is the fraction of the total voltage u that drops across the i-th barrier:

$$\sum_{i=1}^{n+1} \alpha_i = 1 \qquad [20]$$

In order to calculate the electric conductance, Λ, of the single channel in the vicinity of the equilibrium voltage V_e, we assume that a small voltage increment ΔV is applied in addition to V_e:

$$V = V_e + \Delta V \qquad [21]$$

If $\Delta I = ze_o\Phi$ is the electric current, the single-channel conductance is defined by

$$\Lambda = \left(\frac{\Delta I}{\Delta V}\right)_{\Delta V \approx 0} \qquad [22]$$

For small values of the voltage increment ($|e_o\Delta V/kT| \ll 1$), Λ is obtained in the form (see Appendix):

$$\frac{1}{\Lambda} = \frac{kT}{z^2 e_o^2} \cdot \sum_{i=1}^{n+1} \frac{1}{F_i} \qquad [23]$$

Thus, Λ is related in a simple way to the unidirectional fluxes F_i at equilibrium. Equation [23] also shows that for a sequence of barriers with unequal heights, the barrier with the lowest F_i tends to dominate the conductance of the channel. Furthermore, it is interesting to note that Λ is independent of the coefficients α_i. Introducing Eqs. [3] to [5] into Eq. [23], Λ may be expressed in terms of the rate constants $\bar{k}_i'$ and $\bar{k}_i''$:

$$\frac{1}{\Lambda} = \frac{kT}{z^2 e_o^2} \cdot \frac{1}{H} \sum_{i=1}^{n+1} \rho_i \qquad [24]$$

$$\rho_i = \bar{k}_1'' \bar{k}_2'' \ldots \bar{k}_{i-1}'' \bar{k}_i' \bar{k}_{i+1}' \ldots \bar{k}_n' \qquad [25]$$
$$(\rho_1 = \bar{k}_1' \bar{k}_2' \ldots \bar{k}_n'; \ \rho_{n+1} = \bar{k}_1'' \bar{k}_2'' \ldots \bar{k}_n'')$$

We finally apply Eq. [24] to the case of symmetrical aqueous solutions ($c' = c'' = c$) where the equilibrium voltage u_e is zero. Denoting the value of Λ for $u_e = 0$ by Λ_o, one finds, after some rearrangement:

$$\Lambda_o = \frac{z^2 e_o^2}{kT} \cdot \frac{vc\tilde{k}_o'}{1 + \sum\limits_{\nu=1}^{n} \tilde{S}_\nu} \qquad [26]$$

Comparison with Eq. [14] shows that the permeability coefficient at zero voltage, P_o, and the ohmic single-channel conductance, Λ_o, are connected by the relation

$$P_o = \frac{kT}{z^2 e_o^2} \cdot \frac{\Lambda_o}{c} \qquad [27]$$

BIIONIC VOLTAGE

If the channel is permeable to two different ion species, A and B, both of valency z, the resulting zero-current voltage, u_o (the so-called biionic voltage) may be calculated from the relation

$$\Phi_A + \Phi_B = 0 \qquad [28]$$

Introducing Φ_A and Φ_B from Eq. [10], a relation of the form of the well-known Goldman-Hodgkin-Katz equation (11,16) is obtained:

$$zu_o = \ln \frac{P_A c_A'' + P_B c_B''}{P_A c_A' + P_B c_B'} \qquad [29]$$

It should be noted, however, that the permeability coefficients P_A and P_B depend on voltage (Eq. [11]); Eq. [29], therefore, represents an implicit equation for u_o. The reason why we nevertheless retain the form of Eq. [29] is that for certain potential energy profiles the permeability coefficient is only weakly voltage dependent (see below).

REGULAR ENERGY PROFILE

In the following we consider as a special case a channel that consists of $(n + 1)$ regularly spaced barriers of identical height (Fig. 2). This means that $\alpha_i = 1/(n + 1)$ and

$$k_o' = k_a' = k_a e^w; \; k_{n+1}'' = k_a'' = k_a e^{-w} \qquad [30]$$

$$k_\nu' = k_i' = k_i e^w; \; k_\nu'' = k_i'' = k_i e^{-w} \qquad [31]$$
$$(\nu = 1, 2, \ldots, n)$$

$$w \equiv zu/2(n + 1) \qquad [32]$$

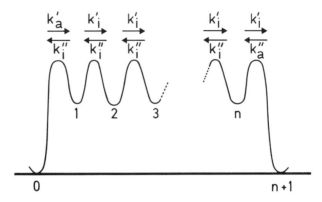

FIG. 2. Channel consisting of $(n + 1)$ regularly spaced barriers of identical height.

k_a and k_i are the rate constants at zero voltage for jumps from the aqueous phase into the channel and for jumps over the internal barriers, respectively. Introduction of Eqs. [30] to [32] into Eqs. [11] and [12] gives

$$P = vk_a \frac{\sinh\left[zu/2(n+1)\right]}{zu/2} \qquad [33]$$

If n is large, P becomes independent of voltage:

$$P \approx \frac{vk_a}{n+1} \qquad [34]$$

It is useful to write this equation in a slightly different form, introducing the equilibrium constant $K = vnk_a/k_i$ for the occupancy of the channel (20):

$$P \approx \frac{K\,k_i}{n(n+1)} \qquad [35]$$

If the channel is permeable to two ion species, A and B, the permeability ratio P_A/P_B is given by, according to Eq. [35]:

$$\frac{P_A}{P_B} = \frac{K_A}{K_B} \cdot \frac{k_{i,A}}{k_{i,B}} \qquad [36]$$

Thus, the permeability ratio is the product of a "thermodynamic" term (containing the two equilibrium constants K_A and K_B for ion binding) and of a "kinetic" term (the ratio of the two rate constants $k_{i,A}$ and $k_{i,B}$), meaning that the ion specificity of a channel is, in general, not only determined by equilibrium parameters, but also by kinetic constants (2,13,14).

We may further introduce the diffusion coefficient $D \approx k_i d^2/(n+1)^2$ of the ion within the channel (10) [d is the length of the channel and therefore $d/(n+1)$ the jump length of the ion within the channel]. This gives (with $n \gg 1$):

$$P \approx \frac{KD}{d^2} \qquad [37]$$

Equation [37] is a continuum approximation for the permeability coefficient. If the membrane contains N_m channels per unit area, the number of occupied channels per unit area in the limit of small ion concentration $c' = c'' = c$ is equal to cKN_m. The quantity KN_m/d can, therefore, be identified with the partition coefficient γ of the ion between membrane and the aqueous phase. The overall permeability coefficient P_m, which is referred to unit area of the membrane, is then obtained in the familiar form

$$P_m = N_m P_i = \frac{\gamma D}{d} \qquad [38]$$

DIFFUSION-LIMITED TRANSPORT RATE OF A CHANNEL

If the intrinsic permeability of a channel is high, the ion flow becomes limited by the rate by which ions from the aqueous solution arrive at the mouth of the channel (12,22,26). The flux through the channel leads to a depletion of permeable ions on one side and to an accumulation on the other side. This effect, commonly called diffusion polarization, diminishes the driving forces acting on the permeating ions. Diffusion-limited ion flow may be treated by considering the channel entrance as a hemispherical sink of radius r_o, the so-called capture radius. In the limit of infinite intrinsic permeability, the ion concentration at the surface of the hemisphere drops to zero and the limiting diffusional flux toward the channel becomes

$$\Phi_{\max} = 2\pi r_o D_w c \qquad [39]$$

c is the ion concentration in the bulk solution and D_w the diffusion coefficient of the ion in water. Equation [39] is strictly valid only if electric potential gradients in the solution adjacent to the channel mouth are negligible. For a more rigorous treatment, both concentration gradients and electric potential gradients have to be taken into account (22). The results may be conveniently expressed by introducing the "convergence permeability," P_c, of the channel:

$$P_c = 2\pi r_o D_w \qquad [40]$$

(For simplicity, we assume the capture radius r_o is the same on both ends of the channel.) If the aqueous phase contains a single $1:1$ electrolyte with permeable cation, the maximum ion flux reached in the limit of infinite intrinsic permeability of the channel is given by (22):

$$\Phi_{\max} = 2P_c \frac{\sqrt{c'}\exp(u/2) - \sqrt{c''}}{\dfrac{1}{\sqrt{c'}}\exp(u/2) + \dfrac{1}{\sqrt{c''}}} \qquad [41]$$

For identical electrolyte concentrations ($c' = c'' = c$), Eq. [41] reduces to

$$\Phi_{max} = 2P_c c \tanh (u/4) \tag{42}$$

This means that the diffusion-limited current saturates at voltages of the order of $u \simeq 4$, or $V \simeq 100$ mV.

If, in addition to the permeable ion species, a large excess of an inert electrolyte is present, Φ_{max} is found to be (22):

$$\Phi_{max} = P_c \frac{c' \exp(u) - c''}{\exp(u) + 1} \tag{43}$$

In a symmetrical system ($c' = c'' = c$), this equation becomes

$$\Phi_{max} = P_c c \cdot \tanh(u/2) \tag{44}$$

In this case where the electric potential difference across the diffusion zone vanishes as a consequence of the high ionic strength, the limiting current for $u \to \infty$ is half the limiting current predicted by Eq. [42]. For $u = 0$ and $c'' = 0$, Eq. [43] becomes $\Phi_{max} = P_c c/2$; this value is only one-half of the flux toward a hemispherical sink (Eqs. [39] and [40]), because either end of the channel acts as a permeation resistance. For $r_o = 0.1$ nm and $D = 10^{-5}$ cm^{2}s^{-1}, P_c becomes approximately $6 \cdot 10^{13}$ cm^3 s^{-1}. The corresponding value of the "convergence conductance" $\Lambda_c = (e_o^2/kT)cP_c$ (compare Eq. [27]) at 1-M concentration c of the permeable ion would be $3 \cdot 10^{-9}$ $\Omega^{-1} = 3$ nS. Accordingly, the range of single-channel conductances Λ for which the diffusion resistance in water is negligible is given by

$$\frac{\Lambda}{c} \ll 3 \text{ nS M}^{-1} \tag{45}$$

For most channels studied so far, Λ/c lies below 3 nS M^{-1}. For instance, recent estimates for the density of sodium channels in squid axon yield $\Lambda/c \simeq 5$ pS M^{-1} (29). Values of Λ/c between 50 and 200 pS have been reported for the ionic channel opened by acetylcholine at the neuromuscular junction (1,18,24). Determinations of the single-channel conductance of gramicidin A in lipid bilayer membranes give Λ/c values of the order of 50 pS M^{-1} for the alkali ions (3,15).

TRANSPORT NOISE FROM ION CHANNELS

In recent years, it has been realized that valuable formation on ion transport mechanisms in biological membranes may be obtained from electrical noise studies (6,7,25,31). Random fluctuations of membrane current and membrane voltage may originate from a number of different sources. One type of noise that has been intensively studied results from the statistical opening and closing of ion channels in the membrane. The analysis of this "channel noise" yields information on the conductance and mean lifetime of the different states of a channel. A second source of noise that is present even in a permanently open channel is associated with the discrete nature of the individual ion translocation steps in the channel. If the movement of an ion within the channel is visualized as a sequence of jumps over energy barriers, then each jump should contribute with a single current pulse to the total current measured in the external circuit. Fluctuations of the total current, therefore, arise from fluctuations in the number of ions crossing the barriers. Such noise, which may be called "transport noise" (30), contains information on the rates of the single transport steps within the channel (9,21,23).

A rather direct theoretical analysis of the transport noise is possible for the equilibrium state of a channel. In this case fluctuations of current I around $I = 0$ are observed at the equilibrium voltage u_e of the transport system. For equilibrium states, the Nyquist theorem (4,5,19,27) may be applied, which relates the spectral intensity S_I of current fluctuations to the small-signal admittance Y of the transport system:

$$S_I(\omega) = 4\,kT \cdot Re[Y(\omega)] \qquad [46]$$

The angular frequency is ω and Re signifies "real part of." Thus, the problem of calculating the noise spectrum is reduced to the calculation of the frequency-dependent admittance $Y(\omega)$, which in turn is related to the relaxation-time spectrum of the channel. If $\tau_\nu (\nu = 1,2, \ldots n)$ are the n relaxation times of a channel with n binding sites (8) (Fig. 1), the spectral intensity of current noise is obtained in the form (23):

$$S_I(\omega) = 4z^2 e_o^2 N_c \left(\sum_{\nu=1}^{n} \frac{\vartheta_\nu}{1 + \omega^2 \tau_\nu^2} + \sum_{i=1}^{n+1} \alpha_i^2 F_i \right) \qquad [47]$$

The frequency-independent quantities ϑ_ν are expressions containing the parameters α_i and the rate constants k_i' and k_i''. Equation [47] may be interpreted in the following way. For a channel with n internal binding sites, the spectral intensity $S_I(\omega)$ has n dispersion regions (regions where S_I changes steeply with frequency) that are centered at angular frequencies $\omega_\nu = 1/\tau_\nu$. In the limit of both low and high frequencies, the transport noise becomes frequency independent ("white"). For $\omega = 0$, S_I is given by

$$S_I(0) = 4z^2 e_o^2 N_c \left(\sum_{\nu=1}^{n} \vartheta_\nu + \sum_{i=1}^{n+1} \alpha_i^2 F_i \right) = 4\,kTN_c\Lambda \qquad [48]$$

$S_I(o)$ represents the thermal noise from a conductance of magnitude $N_c\Lambda$. On the other hand, at high frequencies, correlations between fluctuations in the fluxes Φ_i vanish so that each barrier in the channel behaves as an independent shot noise source of spectral intensity $(2\,z e_o \alpha_i)^2 F_i$ (21):

$$S_I(\infty) = 4z^2 e_o^2 N_c \sum_{i=1}^{n+1} \alpha_i^2 F_i \qquad [49]$$

As mentioned above, these relations have been derived for the equilibrium state of the channel. Recently, Frehland (9) has shown that the Nyquist theorem may also be applied to stationary nonequilibrium states of ion channels.

APPENDIX

Derivation of Eq. [23] (23)

In the presence of a stationary voltage ΔV, the membrane assumes a steady state in which the net ion fluxes Φ_i over all barriers are equal:

$$\Phi_i = k'_{i-1} p_{i-1} - k''_i p_i \equiv \Phi \qquad [A-1]$$

For small voltages ($|\Delta u| = |e_o \Delta V/kT| \ll 1$), Eqs. [18] and [19] may be written in the form

$$k'_i \approx \bar{k}'_i (1 + \alpha_{i+1} z \Delta u/2) \qquad [A-2]$$

$$k''_i \approx \bar{k}''_i (1 - \alpha_i z \Delta u/2) \qquad [A-3]$$

where $\bar{k}'_i$ and $\bar{k}''_i$ are the values of k'_i and k''_i at the equilibrium voltage u_e. Furthermore, for $|\Delta u| \ll 1$ the p_i remain close to their equilibrium values $\bar{p}_i$, so that we may write $p_i = \bar{p}_i(1 + h_i)$ with $|h_i| \ll 1$. Using Eqs. [3], [23], and [24] and neglecting terms proportional to $h_i \Delta u$, Eq. [A-1] gives

$$\frac{\Phi}{F_i} = h_{i-1} - h_i + \alpha_i z \Delta u \qquad [A-4]$$

$$h_o = h_{n+1} = 0$$

Summation over i yields, together with Eq. [20]:

$$\Phi \sum_{i=1}^{n+1} \frac{1}{F_i} = z \Delta u \qquad [A-5]$$

Introducing the electric current $\Delta I = z e_o \Phi$ and combining Eqs. [A-5] and [22] lead to relation 23 for Λ.

REFERENCES

1. Anderson, C. R., and Stevens, C. F. (1973): Voltage clamp analysis of acetylcholine-produced end-plate current fluctuations at frog neuromuscular junction. *J. Physiol. (Lond.)*, 235:655–691.
2. Armstrong, C. H. (1975): Ionic pores, gates, and gating currents. *Q. Rev. Biophys.*, 7:179–210.
3. Bamberg, E., Noda, K., Gross, E., and Läuger, P. (1976): Single-channel parameters of gramacidin A, B, and C. *Biochim. Biophys. Acta*, 419:223–228.
4. Callen, H. B., and Green, R. F. (1952): On a theorem of irreversible thermodynamics. *Physiol. Rev.*, 86:702–710.
5. Callen, H. B., and Welton, T. A. (1951): Irreversibility and generalized noise. *Physiol. Rev.*, 83:34–40.
6. Conti, F., and Wanke, E. (1975): Channel noise in nerve membranes and lipid bilayers. *Rev. Biophys.*, 8:451–506.
7. DeFelice, L. J. (1978): Fluctuation analysis in neurobiology. *Int. Rev. Neurobiol. (in press)*.
8. Frehland, E., and Läuger, P. (1974): Ion transport through pores: transient phenomena. *J. Theor. Biol.*, 47:189–207.

9. Frehland, E. (1978): Current noise around steady states in discrete transport systems. *Biophys. Chem.,* 8:255–265.
10. Glastone, S., Laider, K. J., and Eyring, H. (1941): *The Theory of Rate Processes,* chapter 9. McGraw-Hill, New York.
11. Goldman, D. E. (1943): Potential, impedance, and rectification in membranes. *J. Gen. Physiol.,* 27:37–60.
12. Hille, B. (1970): Ionic channels in nerve membranes. *Prog. Biophys. Mol. Biol.,* 21:3–32.
13. Hille, B. (1975): Ionic Selectivity, saturation, and block in sodium channels. *J. Gen. Physiol.,* 66:635–560.
14. Hille, B. (1975): Ionic selectivity of Na and K channels of nerve. In: *Membranes, A series of Advances, Vol. 3,* edited by G. Eisenmann, pp. 255–323. Marcel Dekker, New York.
15. Hladky, S. B., and Haydon, D. A. (1972): Ion transfer across lipid membranes in the presence of gramacidin A. I. Studies of the unit conductance channel. *Biochim. Biophys. Acta,* 274:294–312.
16. Hodgkin, A. L., and Katz, B. (1949): The effect of sodium ions on the electrical activity of the giant axon of the squid. *J. Physiol. (Lond.),* 108:37–77.
17. Hope, A. B. (1971): *Ion Transport and Membranes,* section 1.4. Butterworths, London.
18. Katz, B., and Miledi, R. (1972): The statistical nature of the acetylcholine potential and its molecular components. *J. Physiol. (Lond.),* 224:655–669.
19. Kubo, R. (1957): Statistical-mechanical theory of irreversible processes. *J. Physiol. Soc. Jpn.,* 12:570–586.
20. Läuger, P. (1973): Ion transport through pores: a rate-theory analysis. *Biochim. Biophys. Acta,* 311:423–441.
21. Läuger, P. (1975): Shot noise in ion channels. *Biochim. Biophys. Acta,* 413:1–10.
22. Läuger, P. (1976): Diffusion-limited ion flow through pores. *Biochim. Biophys. Acta,* 455:493–509.
23. Läuger, P. (1978): Transport noise in membranes: current and voltage fluctuations at equilibrium. *Biochim. Biophys. Acta, 507:337–349.*
24. Neher, E., and Sackmann, B. (1976): Single-channel currents recorded from membrane of denervated frog muscle fibres. *Nature,* 260:799–802.
25. Neher, E., and Stevens, C. F. (1977): Conductance fluctuations and ionic pores in membranes. *Annu. Rev. Biophys. Bioeng.,* 6:345–381.
26. Neumcke, B. (1975): 1/f Membrane noise generated by diffusion processes in unstirred solution layers. *Biophys. Struc. Mech.,* 1:295–309.
27. Nyquist, H. (1928): Thermal agitation of electric charge in conductors. *Physiol. Rev.,* 32:110–113.
28. Parlin, B., and Eyring, H. (1954): Membrane permeability and electrical potential. In: *Ion Transport Across Membranes,* edited by H. T. Clarke, pp. 103–118. Academic Press, New York.
29. Rojas, E., and Keynes, R. D. (1975): On the relation between displacement currents and activation of the sodium conductance in the squid giant axon. *Philos. Trans. R. Soc. Lond. (Biol.),* 270:459–482.
30. Van Vliet, K. M., and Fasset, J. R. (1965): Fluctuations due to electronic transitions and transport in solids. In: *Fluctuation Phenomena in Solids,* edited by R. E. Burgess. Academic Press, New York.
31. Verveen, A. A., and DeFelice, L. J. (1974): Membrane noise. *Prog. Biophys. Mol. Biol.,* 28:189–265.
32. Woodbury, J. W. (1971): In: *Chemical Dynamics, Papers in Honor of Henry Eyring,* edited by J. Hirschfelder. John Wiley and Sons, Inc., New York.
33. Zwolinsky, B. J., Eyring, H., and Reese, C. E. (1949): Diffusion and Membrane Permeability. *J. Phys. Chem.,* 53:1426–1453.

Membrane Transport Processes, Volume 3,
edited by C. F. Stevens and R. W. Tsien.
Raven Press, New York, © 1979.

Problems in the Interpretation of Membrane Current-Voltage Relations

David Attwell

University Laboratory of Physiology, Oxford, OX1 3PT, England

A fundamental aim of membrane electrophysiology is to interpret experimentally measured current-voltage *(I-V)* relationships in terms of the structure of the membrane. This can then, in principle, provide a molecular explanation for the gross electrophysiological properties of the tissue. One obvious problem is how much information on the membrane structure it is possible to get out of *I-V* relationships and whether a unique model of the membrane properties can be produced from the experimental data. Also important is how well the membrane properties can be studied experimentally, given the limitations imposed by the system being studied.

In discussing these points I concentrate on a Nernst-Planck description of membrane properties, rather than using rate theory. This is no limitation as long as membrane permeation systems obeying the independence principle are being studied, but whereas rate theory has been extended to cover the case of pores with interactions between the ions in them, the extension of Nernst-Planck theory to this case is difficult, requiring the evaluation of the chemical potential for a system of interacting ions. Similarly, in discussing experimentally imposed limitations on the study of membrane properties, I ignore the problems of voltage clamp nonuniformity and series resistance, which have been adequately discussed previously, and concentrate on the effects of ion accumulation and depletion outside the membrane's surfaces. In some cells it is possible that such effects are as important in determining cell function as are time-dependent changes in membrane properties of the gated Hodgkin-Huxley type. Furthermore, in the voltage clamping of multicellular preparations, it is quite possible that ignoring accumulation and depletion of ions produces greater problems of interpretation than does lack of voltage control and uniformity.

HOW NONUNIQUE MUST THE INTERPRETATION OF I-V RELATIONS BE?

In this section it is shown that *in principle* for a membrane obeying the Nernst-Planck electrodiffusion equation (plus certain extra assumptions), the *I-V* relationship can be used to obtain an exact expression for the potential energy barrier profile an ion experiences as it crosses the membrane, but that *in practice* experimental inaccuracies greatly limit the amount of information obtainable from the *I-V* curve. The Nernst-Planck equation is used as a basis for this treatment not because it is likely to be a very accurate description of membrane permeation (indeed there is evidence suggesting that many physiological and artificial permeation systems cannot be treated by a simple Nernst-Planck approach; refs. 12,23,27), but because the general principles that come out of the treatment are probably similar for other theoretical frameworks.[1]

The (steady-state) Nernst-Planck electrodiffusion equation treats ion permeation across a membrane as movement across a continuous phase (not necessarily homogeneous) that results from two influences—the concentration gradient dc/dx and the force acting on the ions dU/dx, where c is the ion concentration, U the total energy of the ion, and x the position in the membrane (assumed homogeneous in the plane of the membrane). We then have the expression:

$$I = zFD\{dc/dx + c(zF/RT)\,dU/dx\} \qquad [1]$$

for current across unit area of the membrane. In this equation, z is the ion charge, F is the Faraday, R is the gas constant, T is the absolute temperature, and D is the diffusion coefficient, which can, in general, be position dependent. The use of the steady-state Nernst-Planck equation is valid for time scales of physiological importance since ion redistribution within a membrane is complete after about 10^{-10} sec (9). In the rest of this section, it is assumed that movement of ions from the bulk solutions into the membrane takes a negligible time compared to movement across the membrane; thus interfacial transport is not rate limiting. The dU/dx term in Eq. [1] is the sum of two factors, as the total energy of an ion is due partly to the structure of the membrane and partly to the applied field. Early treatments of ion permeation ignored the membrane contribution to the force and included just the contribution from the applied field (e.g., 8,16,26). However, more recent theoretical work (e.g., ref. 18, and all of the rate-theory treatments) recognizes that the membrane potential energy profile in the absence of an applied field, $U_m(x)$, is unlikely to be a square barrier (leading to $dU_m/dx = 0$). Furthermore, the concept of a localized selectivity filter (11,22) in which a binding site at one point in a pore produces a much greater difference in the energies of ions of different species at that point than at other parts of the pore, implicitly assumes a strongly position-dependent $U_m(x)$. Equation [1] can be integrated without assuming a constant $U_m(x)$ to give (20):

$$\frac{I}{zF} = \frac{C_o \exp\,(-zFV/RT) - C_i}{\displaystyle\int_0^w \frac{\exp\{zFU(x)/RT\}}{D(x)}\,dx} \qquad [2]$$

---

[1] Note that, in the following, "gating" reactions are ignored; it is the interpretation of "instantaneous" *I-V* curves (or equivalently the *I-V* curves of open channels) that is being discussed. Furthermore, only one ion species is assumed permeant; the interpretation of data from systems where more than one species are permeant is discussed later in the section on the Goldman-Hodgkin-Katz equation.

where C_i and C_o are the internal and external ion concentrations, x is the position in the membrane measured from the inside, and the membrane width is w. V is the potential between the bulk solutions $(V = V_i - V_o)$. $U(x)$ is now the potential in the membrane measured with respect to V_i (see ref. 5 for a more detailed description of this derivation). The quantity

$$A = \frac{C_o \exp{(-zFV/RT)} - C_i}{I(V)}$$ [3]

is experimentally measurable. The basic problem of interpreting the *I-V* relation in terms of the membrane potential energy barrier is thus to invert the integral equation

$$A = \int_0^w \frac{\exp{\{zFU(x)/RT\}}}{D(x)} dx$$ [4]

This can be done, with certain assumptions, as follows. In general we can represent the function $G(x) = \exp{\{zFU(x)/RT\}}/D(x)$ as an infinite sum of Legendre polynomials P_n, i.e.,

$$G(x) = \frac{\exp{\{zFU(x)/RT\}}}{D(x)} = \sum_{n=0}^{\infty} a_n P_n \left(\frac{2x}{w} - 1\right)$$ [5]

where the argument $2x/w - 1$ is the distance from the center of the membrane and the coefficients a_n do not depend on x (although they do depend on voltage). This expansion in terms of the orthogonal functions P_n is formally similar to the description of any arbitrary periodic function as an infinite sum of sine and cosine functions, i.e., a Fourier series, and can be done quite generally for any functions $U(x)$ and $D(x)$ that are well behaved. The Legendre polynomials for low n are shown in Fig. 1. So far, this is just a formal restatement of the problem. Without knowing the coefficients a_n, the function $U(x)$ is not known. However, if we assume:

  1. $U(x)$ can be written as the sum of two terms: $U(x) = U_m(x) + f(x, V)$, where $U_m(x)$ is the potential due to the membrane structure and $f(x, V)$ is due to the applied field (making U_m independent of V implies there are no voltage-dependent changes in membrane structure) and
  2. The potential due to the applied V varies linearly across the membrane, i.e., $f(x, V) = -Vx/w$ (this constant superimposed field assumption is only true in the limit of very low charge density in the membrane)

then the coefficients a_n can be determined experimentally (in principle). Thus, the function $\exp{\{zF[U_m(x) - Vx/w]/RT\}}/D(x)$ can, in principle, be found uniquely from the measured *I-V* relationship (5).

The coefficients a_n turn out to be expressible in terms of the differentials of the *I-V* relation as

$$a_n = \sum_{r=0}^{n} b_r \frac{d^r I}{dV^r}$$ [6]

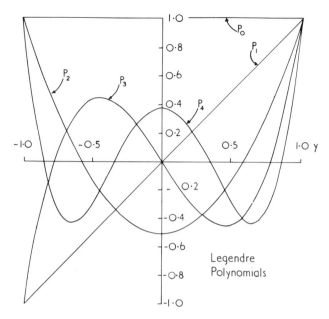

FIG. 1. The Legendre polynomials $P_n(y)$ for $n \leqslant 4$. The analysis method outlined in the text uses an infinite series of these polynomials to obtain an approximation to (the exponential of) the energy profile an ion experiences as it crosses the membrane. Each term in the series is weighted according to how "similar" the exponential of the energy profile is to the Legendre polynomial comprising that term (see Eqs. [5] and [6]).

where the coefficients b_r are known functions of voltage and concentration. This gives useful insight into the accuracy with which the *I-V* relationship has to be measured in order to study the membrane structure to a given level of detail. To include the Nth Legendre polynomial in Eq. [5], the coefficient a_n and hence, from Eq. [6], the Nth differential of the *I-V* relationship must be confidently measurable. We see from Fig. 1 that the Nth Legendre polynomial has N zeroes, so that to resolve detail in the function

$$\exp\{zF[U_m(x) - Vx/w]/RT\}/D(x)$$

on the scale of w/N, one must, roughly speaking, be able to define accurately at least up to the Nth differential of the *I-V* relation.

In principle, all the (infinite number of) terms in the expansion in Eq. [5] must be included to give the function $G(x)$, and this implies knowing the *exact* shape of the *I-V* curve, so that all the differentials (to infinity) can be defined. In practice, however, a limited number of terms can give a reasonable approximation to the function $G(x)$. Examples of this convergence with increasing numbers of terms in the series [i.e., increasing numbers of differentials of $I(V)$ included] are shown in Fig. 2. Here the *I-V* relation derived from Eq. [2] for the $U_m(x)$ consisting of the pair of rectangular barriers (lefthand side) or the single rectangular barrier (righthand side) was used via Eqs. [5] and [6] to obtain successive approximations to $U_m(x)$. $D(x)$ was assumed constant. The number by each

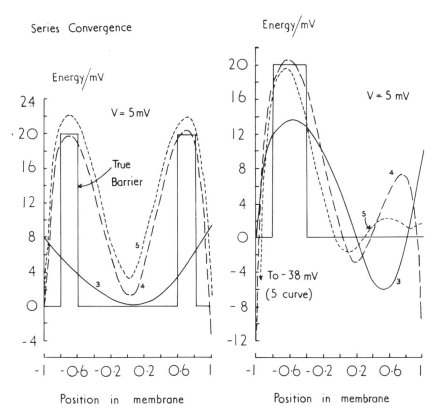

FIG. 2. For the membrane potential energy barriers $U_m(x)$ shown by the continuous rectangular profiles in these two examples, the curves labeled by numbers show successive approximations to the $U_m(x)$, derived using the method in the text. The numbers denote the highest differential of the *I-V* relation used in the series [5].

curve is the highest differential of $I(V)$ considered in each case.

Although, in principle, this is an exact method for obtaining the potential energy barrier from the *I-V* relationship, it is clear that its experimental usefulness will be severely limited by the accuracy of *I-V* data normally available. Experimentally one would probably not place much confidence in the accuracy of above the third differential of an *I-V* relationship, so that one is limited to resolving only the crudest details of the membrane structure. Furthermore, although this inversion of the *I-V* relationship to give $U_m(x)$ has only been derived under rather restrictive assumptions, it seems unlikely that other assumptions will produce very different conclusions about the uniqueness of interpretation of *I-V* data. Consequently, one must expect *in practice* a certain lack of uniqueness in the models of membrane structure one can propose to explain experimental *I-V* relations, at least if one does not consider other data, such as permeability of the system to different ions (cf. ref. 22).

Another aspect of this uniqueness problem is exemplified by the question,

How does one know whether a measured *I-V* relationship is consistent with a Nernst-Planck treatment at all? Some criteria for consistency, and consequently limitations on the types of *I-V* curves Nernst-Planck systems can produce, come directly from the type of analysis presented above (5). For example, the function A must be independent of concentration (eq. [4]; see ref. 18 for the application of this criterion to experimental data). Note that the surface potential must stay constant as the concentration is varied to apply this condition.

If one also makes the constant (superimposed) field assumption, i.e.,

$$f(x) = -Vx/w, \text{ then one can show}$$

$$(-z)^n \frac{d^nA}{dV^n} > 0 \text{ for all } n \tag{9}$$

For $n = 1$, this reduces to $dI/dV > 0$, so that Nernst-Planck systems (with constant field) cannot produce negative slope *I-V* curves of the type found in cardiac and skeletal muscle.

Unfortunately, although the conditions in Eq. [9] must be obeyed for all Nernst-Planck systems (with constant field), it can be shown that Eq. [9] is not sufficient to define this entire class of Nernst-Planck *I-V* relations, i.e., some *I-V* relations can satisfy Eq. [9], but still not be consistent with the Nernst-Planck equation. One approach to check for consistency in this case is to evaluate $G(x)$ from Eqs. [5] and [6] and see if it is positive for all x, as it must be if $U_m(x)$ is to be real. Unfortunately, experimental accuracy again limits such an approach. Although taking all the terms in Eq. [6] will always give a positive sum for $G(x)$ if the system obeys the Nernst-Planck equation with constant field, if the *I-V* accuracy is such that only a few terms can be taken in the series, then the resulting crude approximation to $G(x)$ can be negative. This is especially likely to occur if $U_m(x)$ *varies* by more than RT across the membrane. (The *absolute* value of U_m does not affect this.)

In summary, then, this discussion of how to obtain the membrane potential barrier from the *I-V* relation, under certain rather restrictive assumptions, carries the broader implications that, in general, experimental inaccuracies not only will limit the information that can be obtained from a given *I-V* relation under certain assumptions, but also may not give a definite answer about whether the *I-V* relation is in fact consistent with the assumptions used in analyzing it.

LIMITATIONS ON THE USE OF THE GOLDMAN (ZERO-CURRENT) EQUATION

The Goldman equation (16,26) was originally derived by assuming that the potential energy, $U_m(x)$, of an ion in a membrane, in the absence of an applied field, is independent of position in the membrane and, furthermore, that the applied potential superimposes a constant field on this barrier. Although the constant superimposed field assumption was later relaxed by Ciani et al. (8), the membrane was still assumed to be homogeneous [$U_m(x) = $ constant]. It seems likely, however, that in real membranes $U_m(x)$ varies significantly with position, and it is important to investigate the theoretical validity of the Goldman equation in this case, since it is widely used to characterize the selectivity of physiological membranes permeable to several ion species.

Using the Nernst-Planck equation, with the constant superimposed field assumption, and assuming that interfacial transport of ions is not rate limiting, Attwell and Jack (5) have shown that for the Goldman equation (with constant permeability ratios) to be valid, very specific relationships must hold between the $U_m(x)$ for the different ion species passing through the pore. For two positive monovalent ions, e.g., Na^+ and K^+, the relationship

$$U_{m_{Na}}(x) = U_{m_K}(x) + constant$$

must be obeyed. [Hille (24) has also given this result.] Note that here, for simplicity, it is assumed that the diffusion coefficients are position independent. A more general expression of the above condition is

$$U_{m_{Na}}(x) = U_{m_K}(x) + (RT/F)\text{Log}_e \{D_{Na}(x)/D_k(x)\} + constant$$

Similarly, for positive (K^+) and negative (Cl^-) univalent species in the membrane, the necessary condition for satisfaction of the Goldman equation is $U_{m_K}(x) = -U_{m_{Cl}}(1 - x) + constant$ (where $x = 0$ and $x = 1$ are the membrane edges). Equally restrictive conditions apply for ions of other charges passing through the membrane (5).

If these conditions are not obeyed, the permeability ratios derived from the Goldman equation are voltage dependent. For example, a pore that has a localized selectivity filter at one point in the membrane (see Fig. 3), where the energy of different ion species is very different (i.e., the difference is much greater than at other parts of the membrane), can have a permeability ratio that varies by a factor of 2 over the reversal potential range -75 mV to $+75$ mV.

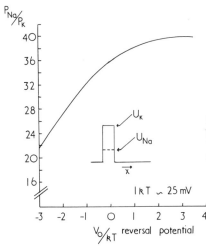

FIG. 3. Variation of P_{Na}/P_K (as defined by the Goldman zero-current equation) with reversal potential. The energy barriers to Na^+ and K^+ shown were chosen to simulate a *localized* selectivity filter.

Clearly, most real membranes' $U_m(x)$ to different ions will not satisfy the conditions necessary for the validity of the Goldman equation with constant permeability ratios. Thus, it seems to be inconsistent to use this equation to evaluate the relative permeability of a membrane to different ions and then to interpret the resulting permeability ratios in terms of *localized* selectivity filters in the membrane. Conversely, if a membrane conductance is found to give constant permeability ratios in the Goldman equation, one might interpret this in terms of all of the pore length setting the selectivity (rather than a localized site). Mullins (36) has suggested a possible mechanism for such a delocalized determinant of selectivity in which different ions have to dehydrate to different extents to pass through the pore, and, furthermore, this dehydration state has to be maintained across all of the membrane.

It is interesting to note that pores showing saturation and block and also transfer mechanisms where interfacial reactions are rate limiting show voltage-dependent permeability ratios if the Goldman equation is used to describe their reversal potentials (8,33). Clearly, voltage dependence of these ratios alone cannot be used to distinguish these types of permeation mechanisms from Nernst-Planck systems where the restrictive conditions mentioned above do not hold.

IMPORTANCE OF ION ACCUMULATION AND DEPLETION FOR I-V RELATIONS

It is often assumed that the ion concentration near the membrane is set by the concentration in the bulk solution being used to perfuse the preparation. This is a necessary assumption, for example, if one wishes to interpret voltage-dependent changes in membrane current in terms of Hodgkin-Huxley gating systems, without confusion arising from the effects on the current of time-dependent changes in the external concentrations. However, in many cases this assumption is not justified because the flow of the current being studied can produce changes in the ion concentration outside the membrane if diffusive contact with the bulk solutions is not good enough. I consider below three different aspects of this problem.

1. First, if the current is distributed homogeneously in the plane of the membrane and if there is no physical barrier to diffusion outside the membrane, the presence of an unstirred layer of solution near the membrane can result in changes of concentration (from the bulk solution values) near the membrane. Läuger and Neumcke (34) have treated this situation and, assuming an unstirred layer of thickness 0.02 cm, have concluded that the condition:

$$\frac{\text{membrane conductance}}{\text{ion concentration in bulk solution}} = \frac{\lambda}{c} \ll 0.5 \; \Omega^{-1} \, cm^{-2} \, M^{-1}$$

must be satisfied if the steady-state current is not to significantly affect the ion concentration outside the membrane. (Strictly speaking, λ should be the conductance in the absence of diffusion polarization). Although the thickness of unstirred layer chosen by Läuger and Neumcke may not be appropriate to all preparations, they point out that for many lipid bilayer experiments the condition above is not satisfied and the so-called diffusion polarization can dominate the properties of the measured current. If we follow Läuger and Neumcke and assume an unstirred layer thickness of 0.02 cm for the squid axon (although here a more important feature may be the presence of a physical barrier outside the cell membrane; see later), the measured sodium current conductance of $\sim$ 100 mmho/cm^2 (25) and external [Na$^+$] of 400 mM gives $\lambda/c \sim 0.25 \ \Omega^{-1}$ cm^{-2} M^{-1} so that the condition above is not very well satisfied. Similarly, for the sodium current in the myelinated node, the values for the current quoted by Bergman (7) lead to a value for λ/c of $\sim 1 \ \Omega^{-1}$ cm^{-2} M^{-1}. Again there is an additional physical diffusion barrier outside the nodel membrane in this preparation (10).

In cardiac muscle, on the other hand, Beeler and Reuter (6) found a peak g_{Na} of about 200 Ω^{-1} cm^{-2} so that

$$\frac{\lambda}{c} = 0.05 \ \Omega^{-1} \ cm^{-2} \ M^{-1} \ll 0.5 \ \Omega^{-1} \ cm^{-2} \ M^{-1}$$

and diffusion polarization may well be unimportant.

2. Second, in the rough calculations above, it is assumed that the current is spread uniformly in the plane of the membrane. If the current is localized in pores, however, the resulting change in ion concentration near the mouth of the pore is larger than if the same current were distributed uniformly over all the membrane area. To obtain a rough quantitative estimate of this effect, let us consider one pore (ignoring the rest) and treat the ions that flow through the pore as diffusing, on leaving the pore, into an infinite hemisphere of solution. The mouth of the pore is treated as a hemisphere of radius "a." At infinity the concentration has the bulk solution value, but nearer the mouth of the pore the concentration is higher, for a net ion flux toward that side of the membrane (lower, for a net flux away from that side of the membrane). Solving the diffusion equation

$$\frac{I}{F} = D \, 2\pi r^2 \frac{dc}{dr}$$

for a steady-state current I of ions with diffusion coefficient, D, and concentration, c, at radius, r, from the pore mouth, we find the difference in concentration, Δc, between the mouth of the pore and infinity is

$$\Delta c = \frac{I}{2\pi DFa}$$

where F is the Faraday. Note that here I assume a totally unstirred phase, so that Δc is the concentration difference from $r = a$ to $r =$ infinity. However if one assumes that at a radius b from the pore, the solution is well stirred, then so long as b $\gg$ a (e.g., b $\sim$ 0.02 cm as Läuger and Neumcke assumed), the above expression is a very good approximation. Taking the conductance of one sodium pore as 2.5×10^{-12} mho (29), a driving force of 100 mV, $D = 10^{-5}$ cm^{-2} sec^{-1}, and $a = 3$Å, we find $\Delta c \sim 0.7$ mM. This is an accumulation on one side of the membrane, and a depletion on the other side. Clearly this change in concentration is negligible compared to the normal intra- (or extra-) cellular sodium concentration.

However, this effect may be significant for the case of calcium pores. The conductance of a calcium pore is unknown; however, on a Nernst-Planck treatment if the driving force for a current is strongly inward, the current is proportional to the external concentration. This factor, alone, will make the current through a calcium pore be a factor of 200 smaller than that through a sodium pore (for external concentrations of 2 and 400 mM, respectively). Neglecting the obvious differences in pore structure and the charge on the ion, one might thus guess the conductance of a calcium pore to be of the order of 1/200th of that of a sodium pore. The resulting value for Δc is 0.003 mM [neglecting the possibility of [calcium] buffering near the mouth of the pore]. This is small in absolute magnitude, but not compared to [Ca^{2+}]$_i$, which is about 10^{-8} to 10^{-7}M (larger in the Debye-Hückel layer if the membrane has a negative surface charge). Thus the flow of current and concomitant diffusion of ions away from the mouth of the pore (remembering that inside the cell the current will be largely carried by K$^+$) could significantly affect the reversal potential of this current. If this speculation is correct, then the time course of the current will also be affected by the diffusion polarization, and this could contribute to the observed inactivation of the calcium current (39).

3. Third, in the calculations above, the cause of the change in ion concentration near the membrane has been simply the time needed for ions passing through the membrane to diffuse into a semiinfinite solution. In some preparations, the assumption of a semiinfinite solution may not be a good one. For example, at the myelinated node the geometry of the axon is such that the change of concentration near the nodal membrane is significantly greater than it would be if the inside of the membrane faced a semiinfinite solution (2,4,7). (Note that this is quite apart from the barrier to diffusion outside the nodal membrane, ref. 10.) Similar situations of restricted geometry may exist in the cell bodies and dendritic trees of neurons (e.g., dendritic spines) and could play an important role in determining the information-processing properties of such cells (17,19,38).

Glial cells often provide a restriction to diffusion to the outside of neuronal membranes. This has been best studied in squid giant axon (see refs. 1 and 15) and myelinated node (see ref. 10) preparations, but is probably important in many central nervous system cells since extracellular accumulation of potassium ions has been shown to occur in the spinal cord (31), medulla (32), hippocampus (14), and cerebral cortex (21). Similarly, many cardiac and smooth muscle preparations have an endothelial layer around them, restricting diffusion from the bulk solution to the space outside the cell membranes, or have a tortuous extracellular cleft structure, again reducing diffusive contact between the cell membrane and the bulk solution. Large changes in the concentration of K^+ and Ca^{2+} in the extracellular spaces can occur (13,30,35).

Ion accumulation and depletion probably have a significant effect on the time course of membrane currents in many preparations, but at present, there is no rigorous theoretical framework available for separating such effects from changes in current due to changes in Hodgkin-Huxley-type gating variables. However, by analyzing a grossly oversimplified model (13), in which one considers a membrane to be in good diffusive contact with the bulk solution on one side of it and separated from the bulk solution on the other side by a lumped extracellular space and a simple lumped diffusion barrier (see Fig. 4), it is possible to gain some insight into the properties conferred on a membrane by the occurrence of ion accumulation and depletion. If the diffusion barrier between the bulk solution and the extracellular space is severe, the dependence of the membrane current on the (experimentally adjustable) bulk solution concentration can be very different from the "true" dependence on the ion concentration just outside the membrane. Interestingly, however, on this simple model, at least, it turns out that if the *I-V* relationships at different bulk ion concentrations crossover (e.g., ref. 37), then the *I-V* relationships at different fixed ion concentrations just outside the membrane must also crossover. Thus, the presence of a restricted extracellular space cannot generate artifactual crossovers in *I-V* relationships. Another interesting conclusion from the analysis of such a model is that the presence of a restricted extracellular space alters the stability conditions for the membrane potential of the cell. Whereas a patch membrane in the absence of such a restricted space has a zero-current potential stable to small displacements from its equilibrium value if the slope conductance is positive (e.g., ref.

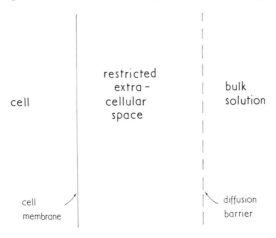

FIG. 4. A grossly oversimplified model of a preparation with a restricted extracellular space. In most physiological preparations, the space will be composed of tortuous extracellular pathways and will not be well mixed. However, for theoretical analysis it is convenient to lump the extracellular space into one well-mixed compartment, separated from the bulk solution by a lumped diffusion barrier.

28), this is not necessarily true in the presence of a restricted space. A zero-current potential can be stable even if the slope conductance is negative and can be unstable even if the slope conductance is positive (3). Such effects may be important in relation to the changing of the threshold of neuronal firing, when a cell is undergoing repetitive activity.

ACKNOWLEDGMENTS

Much of this work has been done in conjunction with Julian Jack, David Eisner, and Ira Cohen. I am grateful to Laurence Waters for photography, Gill Smith for drawing, and Sarah Snell for typing. During this work I was a Medical Research Council scholar.

REFERENCES

1. Adelman, W. J., Palti, Y., and Senft, J. P. (1973): Potassium ion accumulation in a peri-axonal space, and its effects on the measurement of potassium ion conductance. *J. Membr. Biol.,* 13:387–410.
2. Århem, P. (1976): Diffusion of sodium in axoplasm of myelinated nerve fibre. *Acta Physiol. Scand.,* 97:415–425.
3. Attwell, D., Cohen, I., and Eisner, D. A. (1978): Membrane potential stability conditions for a cell with a restricted extra-cellular space: A role for restricted spaces in the initiation of arrhythmias? *(In preparation.)*
4. Attwell, D., and Jack, J. J. B. (1978): Ion accumulation at the myelinated node. *(In preparation.)*
5. Attwell, D., and Jack, J. J. B. (1978): The interpretation of membrane current-voltage relations: a Nernst-Planck analysis. *Prog. Biophys. Mol. Biol. (in press.)*
6. Beeler, G. W., and Reuter, H. (1970): Voltage clamp experiments on ventricular myocardial fibres. *J. Physiol. (Lond.),* 207:165–190.
7. Bergman, C. (1970): Increase of sodium concentration near the inner surface of nodal membrane. *Pfluegers Arch.,* 317:287–302.
8. Ciani, S. M., Eisenman, G., and Szabo, G. (1969): A theory for the effects of neutral carriers such as the macrotetralide actin antibiotics on the electrical properties of bilayer membranes. *J. Membr. Biol.,* 1:1–36.
9. Cole, K. C. (1968): *Membranes, Ions and Impulses.* Univ. of California Press, Berkeley and Los Angeles.
10. Dubois, J.-M., and Bergman, C. (1975): Potassium accumulation in the perinodal space of frog myelinated axons. *Pfluegers Arch.,* 358:111–124.
11. Eisenman, G. (1962): Cation selective glass electrodes and their mode of operation. *Biophys. J.,* 2 (no. 2, part 2): 259–323.
12. Eisenman, G., Sandblom, J., and Neher, E. (1976): Evidence for multiple occupancy of Gramicidin A channels by ions. *Biophys. J.,* 16:81a (Abstr.)
13. Eisner, D. A., Cohen, I., and Attwell, D. (1978): The interpretation of voltage clamp and tracer flux data in preparations with a restricted extra-cellular space. *(In preparation.)*
14. Fertiziger, A. P., and Ranck, J. B. (1970): Potassium accumulation in interstitial space during epileptiform seizures. *Exp. Neurol.,* 26:571–585.
15. Frankenhaeuser, B., and Hodgkin, A. L. (1956): The after effects of impulses in the giant nerve fibres of Loligo. *J. Physiol. (Lond.)* 131:341–376.
16. Goldman, D. E. (1943): Potential, impedance and rectification in membranes. *J. Gen. Physiol.,* 27:37–60.
17. Grossman, Y., Spira, M. E., and Parnas, I. (1973): Differential flow of information into branches of a single axon. *Brain Res.,* 64:379–386.
18. Hall, J., Mead, C. A., and Szabo, G. (1973): A barrier model for current flow in lipid bilayer membranes. *J. Membr. Biol.,* 11:75–97.

19. Hatt, H., and Smith, D. O. (1975): Axon conduction block: Differential channeling of nerve impulses in the crayfish. *Brain Res.,* 87:85–88.
20. Haydon, D. A., and Hladky, S. B. (1972): Ion transport across thin lipid membranes: A critical discussion of mechanisms in selected systems. *Q. Rev. Biophys.,* 5:187–282.
21. Heinemann, U., and Lux, H. D. (1975): Undershoots following stimulus-induced rises of extracellular potassium concentration in cerebral cortex of cat. *Brain Res.,* 93:63–76.
22. Hille, B. (1972): The permeability of the sodium channel to metal cations in myelinated nerve. *J. Gen. Physiol.,* 59:637–658.
23. Hille, B. (1975): Ionic selectivity, saturation and block in sodium channels. A four barrier model. *J. Gen. Physiol.,* 66:535–560.
24. Hille, B. (1976): Ionic selectivity of Na and K channels of nerve membranes. In: *Membranes, Vol. 3: Lipid Bilayers and Biological Membranes: Dynamic Properties,* edited by G. Eisenman, Chapt. 4. Dekker, New York.
25. Hodgkin, A. L., and Huxley, A. F. (1952): Membrane current in nerve. *J. Physiol. (Lond.),* 117:500–544.
26. Hodgkin, A. L., and Katz, B. (1949): The effect of sodium ions on the activity of the giant axon of the squid. *J. Physiol. (Lond.),* 108:37–77.
27. Hodgkin, A. L., and Keynes, R. D. (1955): The potassium permeability of a giant nerve fibre. *J. Physiol. (Lond.),* 128:61–88.
28. Jack, J. J. B., Noble, D., and Tsien, R. W. (1975): Electric current flow in excitable cells. Clarendon Press, Oxford.
29. Keynes, R. D., and Rojas, E. (1974): Kinetics and steady state properties of the charged system controlling sodium conductance in the squid giant axon. *J. Physiol. (Lond.),* 239:393–434.
30. Kline, R., and Morad, M. (1976): Potassium efflux and accumulation in heart muscle. Evidence from K$^+$ electrode experiments. *Biophys. J.,* 16:367–372.
31. Kříž, N., Syková, E., Ujec, E., and Vyklický, L. (1974): Changes of extracellular potassium concentration induced by neuronal activity in the spinal cord of the cat. *J. Physiol. (Lond.),* 238:1–15.
32. Krnjevic, K., and Morris, M. E. (1972): Extracellular K$^+$ activity and slow potential changes in spinal cord and medulla. *Can. J. Physiol. Pharmacol.,* 50:1214–1217.
33. Läuger, P. (1973): Ion transport through pores: A rate theory analysis. *Biochim. Biophys. Acta,* 311:423–441.
34. Laüger, P., and Neumcke, B. (1973): Ion conductance in lipid bilayers. In: *Membranes: A Series of Advances,* Vol. II, edited by G. Eisenman; pp. 1–59. Marcell Dekker, New York.
35. Maughan, D. N. (1973): Some effects of prolonged depolarization on membrane currents in bullfrog atrial muscle. *J. Membr. Biol.,* 11:331–352.
36. Mullins, L. J. (1975): Ion selectivity of carriers and channels. *Biophys. J.,* 15:921–931.
37. Noble, D., and Tsien, R. W. (1968): The kinetics and rectifier properties of the slow potassium current in cardiac Purkinje fibres. *J. Physiol. (Lond.),* 220:547–563.
38. Parnas, I. (1972): Differential block at high frequency of branches of a single axon innervating two muscles. *J. Neurophysiol.,* 35:902–914.
39. Reuter, H., and Scholz, H. (1977): A study of the ion selectivity and the kinetic properties of the calcium dependent slow inward current in mammalian cardiac muscle. *J. Physiol. (Lond.),* 264:17–47.

Membrane Transport Processes, Volume 3,
edited by C. F. Stevens and R. W. Tsien.
Raven Press, New York, © 1979.

Chemistry and Ionic Conductivity in Beta and Beta" Alumina

Gregory C. Farrington

General Electric Research and Development Center, Schenectady, New York 12301

COMPOSITION, STRUCTURE, AND CONDUCTIVITY OF ALKALI-SUBSTITUTED BETA AND BETA" ALUMINA

Ionic diffusion and conduction in crystalline solids are two related phenomena that have been the subjects of extensive experimental and theoretical investigation. Particular attention has been focused on conductivity in alkali halides (13). If a crystal of a simple alkali halide such as LiF were perfect with no lattice sites vacant, then ionic conduction could not occur. Diffusion might be observed as the result of two adjacent ions exchanging positions, but net charge transfer, which is the essential process of ionic conductivity, could not take place. Conductivity in crystalline solids requires the presence of vacancies among normally occupied lattice sites or of interstitial sites. Therefore, the nature and mechanism of ionic conductivity in crystalline solids involve broader questions of defects and imperfections in crystal lattices.

Two models of lattice defects fundamental to ionic conductivity result from the work of Schottky (23) and of Frenkel (9). So-called Schottky and Frenkel defects are shown in Fig. 1. Schottky observed that unoccupied lattice sites can occur in a crystal if there are equivalent numbers of anion and cation vacancies. Ion migration can then take place by ionic hopping from filled to adjacent vacant sites. Frenkel, in contrast, proposed that thermal vibrations might be sufficient to promote an ion out of its normal lattice site into a higher energy interstitial position. Conduction can then take place if that ion drops back to a vacant lattice site or cooperatively promotes another ion from its lattice site into a new interstitial site, replacing it in the lattice. These models stimulated a great number of investigations into conductivity and ionic diffusion in simple crystalline compounds both in pure and in doped compositions in which lattice vacancies had been deliberately introduced. With either model, however, it is obvious that for a compound to display high conductivity three criteria must be met: (a) the concentration of potentially mobile ions must be large, (b) the concentration of alternate ionic sites, either vacancy or interstitial, must also be large, and (c) the energy required for an ion to move from a filled to unfilled site must be small. Conduction in simple alkali halide crystals

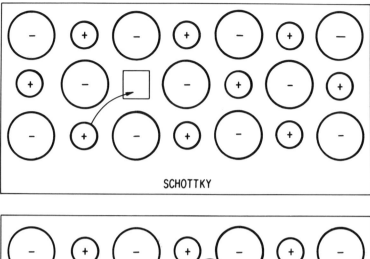

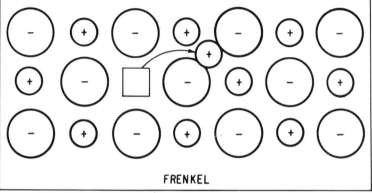

FIG. 1. Schottky and Frenkel defect models.

has now been shown to generally occur by the Schottky mechanism. Conduction can often be described by a simple Arrhenius relationship (Eq. [1], see Table 1 for symbols and units) for activated hopping

$$\sigma T = \sigma_o \exp(-E_a/RT) \qquad [1]$$

from one potential energy minimum to another across an intervening energy barrier, if it is assumed that lattice interactions following the transition rapidly damp the ionic motion. Nearly all alkali halides have conductivities many orders of magnitude smaller than observed in liquid aqueous and nonaqueous electrolytes.

New interest in the area of solid ionic conductors was stimulated by the discovery that the Na^+ conductivity of sodium beta alumina $[(1+x)Na_2O \cdot 11Al_2O_3]$ in single crystal form is about 10^{-2} (ohm cm)$^{-1}$ (33) at 25°C. It varies in a linear Arrhenius relationship from −50 to 800°C with an activation energy of 3.7 kcal/mole (0.17 eV) and is comparable to that of nonaqueous electrolyte solutions, such as $LiClO_4$ in propylene carbonate. Although new solid ionic conductors of comparable conductivity have since been identified (24), sodium beta alumina and a related compound sodium beta″ alumina remain particularly interesting because they serve as host structures for the conduction of a variety of monovalent cations. For example, the entire Na^+ content of sodium beta alumina can be replaced by Li^+, K^+, Ag^+, H_3O^+, H^+, and NH_4^+. Similar ion exchange chemistry occurs with sodium beta″ alumina.

Much of the investigation of conductivity in the beta alumina structure has focused on sodium beta alumina. This chapter first discusses the composition and structure of sodium beta and beta″ alumina, the conductivity of each composition, and its structural basis. Then, recent investigations of composition, structure, and ionic conductivity in H_3O^+ beta and beta″ alumina, Li^+–Na^+ beta and beta″ alumina, and hydration reactions of alkali beta and beta″ aluminas are reviewed. Particular attention is focused on ion–lattice and ion–ion interactions that influence conductivity in the structures.

Crystal Structure of Sodium Beta Alumina

Sodium beta alumina is a nonstoichiometric compound (18) with an ideal formula of $Na_2O \cdot 11\ Al_2O_3$. Its structure was first determined by Bragg (3) and later refined by Beevers and Ross (2), Peters et al. (15), and Roth (18). Peters et al. and Roth et al. particularly addressed the arrangement of sodium ions within the so called "conducting planes." The beta alumina structure has recently been reviewed by Roth et al. (21).

Sodium beta alumina has a hexagonal layer structure with lattice constants of $a_o = 5.59$ Å and $c_o = 22.53$ Å. Sodium ions are found in planes located about 11.23 Å apart, perpendicular to the c axis, which are referred to as the conducting planes. Ionic motion takes place in two dimensions within the conducting planes but not perpendicular to them.

Close-packed blocks of aluminum and oxygen ions, called "spinel blocks" for their resemblance to the $MgAlO_4$ (spinel) structure, separate the conducting planes. Aluminum ions occupy both octahedral and tetrahedral sites in the structure. The conducting planes are bounded by close-packed oxygen layers above and below, about 4.76 Å apart. In the beta alumina structure, the close-packed oxygen layers are eclipsed in relation to each other and are bound together by sodium ions and Al—O—Al bonds with oxygen ions within the conducting plane. The conducting planes are therefore mirror planes. Figure 2 is a view of the sodium beta alumina structure in a direction perpendicular to the c axis.

Long-range ionic motion and conduction occur within the conducting plane, the structure of which is illustrated in Fig. 3. The upper close-packed layer of oxygens is not shown. Sodium ions are distributed among three nonequivalent crystallographic sites, the so-called Beevers–Ross, midoxygen, and anti-Beevers–Ross sites. In Fig. 3, the sodium ions are shown in the Beevers–Ross sites,

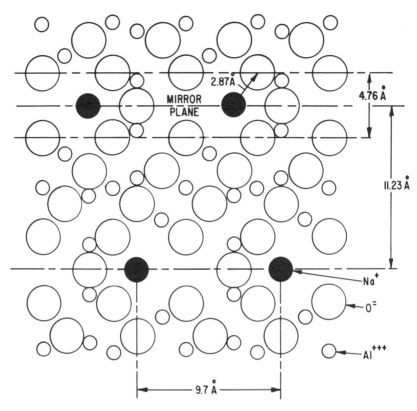

FIG. 2. The crystal structure of beta alumina as seen perpendicular to the c axis. Conduction occurs in the relatively open mirror planes.

those most favorable for ion occupation. A sodium ion in the center of each Beevers–Ross site is 2.88 Å from each of three oxygen ions in the close-packed layer both above and below the conducting plane. The ion is equidistant from three oxygen ions in the conducting plane at a distance of 3.25 Å. In the midoxygen site, a sodium ion is also coordinated to six close-packed oxygen ions, three in each plane at a distance of 2.88 Å, but its nearest neighbors are two oxygen ions in the conducting plane, 2.80 Å away. The anti-Beevers–Ross site, like the Beevers–Ross site, has a sodium ion coordinated with three oxygen ions in the conducting plane, each 3.25 Å away, but placed between two oxygens above and below the conducting plane, 2.38 Å away. The distances cited are from the centers of the sodium ions to the centers of the oxygen ions. The site in which a sodium ion comes closest to neighboring oxygen ions in or bounding the conducting plane is the anti-Beevers–Ross site in which the sodium ion moves between a pair of oxygen ions through a spacing 2.0 Å wide. For a sodium ion, which has a diameter of 1.94 Å, the spacing is still larger than the ion. However, for larger ions such as K^+ or Rb^+, which have diameters

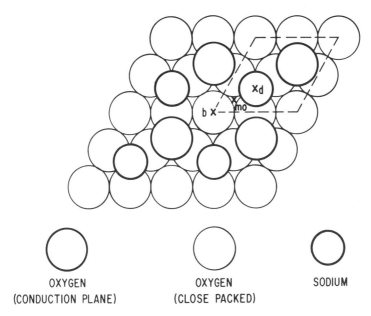

OXYGEN
(CONDUCTION PLANE)

OXYGEN
(CLOSE PACKED)

SODIUM

FIG. 3. Ionic arrangement in the conducting plane of beta alumina. The top layer of close-packed oxygens bounding the plane has been lifted away. The three nonequivalent sites for sodium ion occupation are labeled "d" (Beevers–Ross), "mo" (midoxygen), and "b" (anti-Beevers–Ross). Sodium ions are shown in the Beevers–Ross sites, those most favorable for sodium ion occupation.

of 2.66 and 2.94 Å, respectively, movement through the anti-Beevers–Ross is energetically more difficult.

Peters et al. (15) and Roth et al. (21) have shown that the relative occupancies of the Beevers–Ross, midoxygen, and anti-Beevers–Ross sites vary with temperature. At 25°C, a sodium ion is most likely found in the Beevers–Ross site, less likely in the midoxygen site, and least in the anti-Beevers–Ross site. This distribution becomes more uniform with increasing temperature. At −196°C, sodium ion density is almost exclusively in the Beevers–Ross and midoxygen sites.

The stoichiometric formula for sodium beta alumina is $Na_2O \cdot 11\ Al_2O_3$, but it occurs as a nonstoichiometric compound with a formula more accurately expressed as $(1 + x)Na_2O \cdot 11\ Al_2O_3$. The upper limit for x is not clearly known; in typical compositions x has a value of 0.3. It is not yet known whether stoichiometric sodium beta alumina can be synthesized. If sodium beta alumina were a stoichiometric compound, one sodium ion would be available for each Beevers–Ross site. Nonstoichiometric excess sodium ions also are found in the conducting plane and interact with ions previously in Beevers–Ross sites to form midoxygen ion pairs. It has been proposed (18) that the excess sodium ion content of sodium beta alumina is compensated for by interstitial oxygen ions located in midoxygen sites in the conducting plane.

Crystal Structure of Sodium Beta" Alumina

Bettman and Peters (1) have proposed an ideal composition of sodium beta" alumina of $Na_2O \cdot 5\ Al_2O_3$, although actual compositions contain sizable concentrations of foreign ions such as Mg^{2+} or Li^+ 'to stabilize' the structure. The result is a ternary phase that in one form may be written as $Na_2O \cdot MgO \cdot 5\ Al_2O_3$. Yamaguchi and Suzuki (28), Bettman and Peters (1), and Roth et al. (21) have determined the structure of sodium beta" alumina, which is quite similar to that of sodium beta alumina. In both compounds, the sodium ions are located in planes separated by spinel blocks. In each, the spinel blocks are identical. The key difference between the structures is that neighboring spinel blocks in beta" alumina are rotated $\pi/3$ relative to their configuration in beta alumina. The beta alumina structure is hexagonal, the beta" rhombohedral.

From the standpoint of conductivity, the key difference between beta and beta" alumina is in the structure of the conducting planes. The sodium beta alumina conducting plane is shown in Fig. 3 and has already been discussed. The ionic arrangement in the beta sodium beta" alumina conducting plane is shown in Fig. 4. Whereas three nonequivalent sites for mobile ion occupation are present in sodium beta alumina, all sodium sites are crystallographically equivalent in the sodium beta" alumina structure. Mobile ion sites alternate in beta" alumina between configurations in which three oxygen ions are above

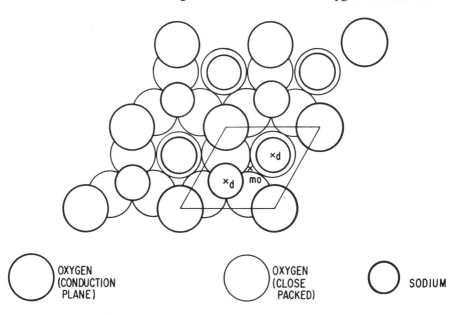

FIG. 4. Ionic arrangement in the conducting plane of beta" alumina. The top layer of close-packed oxygens bounding the plane has been lifted away. When the top oxygen layer is in place, sodium ions alternate between identical "d" sites first above three oxygen and below one and then below three oxygens and above one. The midpoint of an ionic hop is the midoxygen site, mo.

a sodium ion at a distance of 2.69 Å and one below 2.57 Å and in which three oxygen ions are below the sodium ion and one above. In each configuration, a sodium ion is surrounded by three oxygen ions in the conducting plane at a distance of 3.25 Å. Because a sodium ion is alternately 2.57 Å from a single oxygen ion below and then the same distance from an oxygen ion above, it undulates through the conducting plane, occupying positions 0.17 Å either above or below the midpoint of the plane. At the midpoint of an ionic hop from an occupied to an unoccupied site, the center of a sodium ion is 2.8 Å from the centers of two oxygen ions in the conducting plane. The "open" space between the oxygen ions is 3.0 Å and is the smallest space through which a sodium ion must migrate in beta″ alumina. In contrast, with beta alumina a sodium ion must move through a space 2.0 Å wide in the anti-Beevers–Ross site. The diameter of a sodium ion is 1.94 Å, smaller than the narrowest gap in either structure. However, for ions larger than Na^+, such as K^+, Rb^+, and Cs^+, passage through the anti-Beevers–Ross site in the beta alumina structure requires more energy than passage through the midoxygen site in the beta″ alumina structure. This should be reflected in lower activation energies for conduction of these ions in beta″ alumina.

Ionic Conductivity in Beta and Beta″ Alumina

It has been proposed (30) that conductivity in sodium beta alumina occurs by an interstitialcy mechanism. Central to this concept is the existence of midoxygen/midoxygen Na^+ ion pairs in the conducting plane. If sodium beta alumina were stoichiometric, one sodium ion would be available for every Beevers–Ross site. As mentioned previously, sodium beta alumina is nonstoichiometric having about 30% excess sodium ions. Injecting excess sodium ions into the structure involves adding ions to either the anti-Beevers–Ross sites or the midoxygen sites. Neutron diffraction data (21) reveal that the excess ions reside primarily in the midoxygen sites at 25°C or lower. Interionic repulsion between two ions in the region of a Beevers–Ross site produces a midoxygen–midoxygen pair. So, in a simple model, the process of conduction may be depicted as in Fig. 5 in which the movement of a second ion into a singly occupied Beevers–Ross site results in the displacement of the ion initially present, forming midoxygen–midoxygen pair. Consequent diffusion of one ion by further diffusion restores the other to Beevers–Ross site. Sodium ion distribution in beta alumina, then, is the result of ion–lattice interactions on which ion–ion interactions are superimposed.

The results of direct conductivity measurements on small single crystals of sodium beta alumina and alkali ion-exchanged isomorphs reported by Whittingham and Huggins (30) are listed in Table 1. These measurements were carried out by contacting the crystals with reversible electrodes in the form of various alkali bronzes, such as cubic lithium tungsten bronze and sodium vanadium bronze, which are both electronic conductors and ionic conductors for their respective alkali ions.

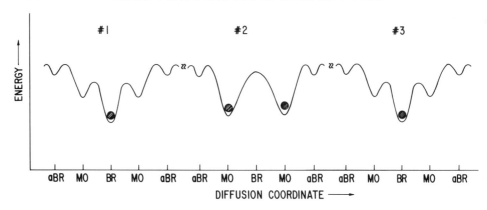

FIG. 5. Qualitative potential well diagram illustrating interstitialcy mechanism of conductivity in beta alumina. The shapes of the potential wells will differ depending on ionic type. aBR, anti-Beevers–Ross; BR, Beevers–Ross; MO, midoxygen.

The conductivity of sodium beta alumina obeys a simple Arrhenius relationship (Eq. [1]) over the remarkably large temperature range of −150 to 800°C (Fig. 6). The activation energy observed, 3.71 kcal/mole or 0.17 eV, is quite small compared to simple alkali halides, for which activation energies are in the range of 10 to 20 kcal/mole. The conductivity of sodium beta alumina at 25°C, 10^{-2} (ohm cm)$^{-1}$, is comparable to that of nonaqueous liquid electrolytes such as $LiClO_4$ (1 M) in propylene carbonate. For K^+ and Rb^+ beta alumina the room temperature conductivities decrease as the result of increasing activation energies. This trend is to be expected as sodium ions are replaced by larger alkali ions for which motion through the anti-Beevers–Ross site requires greater energy. The exception to this trend is lithium beta alumina, which has a lower conductivity than sodium beta alumina at 25°C because of a larger activation energy. It has been suggested (29) that the higher activation for lithium motion occurs because the small lithium ion preferentially associates with one or the other close-packed oxygen layers bounding the conducting plane. Lithium beta alumina is also the only alkali beta alumina that shows a "kink" in its Arrhenius curve,

TABLE 1. *Conductivity of beta alumina isomorphs*

Ion	Temp. range (°C)	E_a (kcal/mole)	σ_0 (ohm-cm)$^{-1}$°K	σ25°C (ohm cm)$^{-1}$	Reference
Na	−150–820	3.71	2.4×10^3	1.4×10^{-2}	(30)
Ag	25–800	3.87	1.6×10^3	6.7×10^{-3}	(30a)
K	−70–820	6.67	1.5×10^3	6.5×10^{-5}	(30a)
Tl	−20–800	8.05	6.8×10^2	2.2×10^{-6}	(30a)
Li	180–800	8.41	9.7×10^3	—	(30a)
H_3O^+	20–200	18.	7.8×10^4	1.7×10^{-11}	(7)
H^+–H_3O^+	300–550	29.	2.5×10^4	—	(7)

From Whittingham & Huggins, ref. 30.

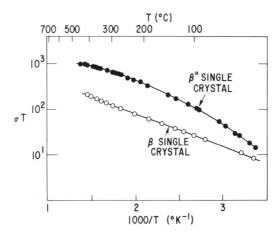

FIG. 6. Arrhenius plot of single crystal conductivity of sodium beta and sodium beta″ alumina. T, temperature. (From refs. 12 and 30.)

the conductivity above 180°C varying with a higher activation energy than below 180°C. If this observation is accurate, then lithium beta alumina is the only alkali beta alumina for which there is clear evidence of two conduction mechanisms.

Dielectric loss measurements (17) on powdered samples of various beta aluminas and tracer diffusion measurements (29) have also been carried out. Because of the high conductivity of alkali beta aluminas, dielectric loss peaks only are observed at readily accessible frequencies at low temperatures (0 to −100°C). Tracer diffusion takes place at practical rates only at elevated temperatures (200 to 500°C). Therefore, data from each of these techniques are only available in restricted temperature ranges. For Na$^+$, K$^+$, and Ag$^+$ beta aluminas, single crystal conductivity measurements, dielectric loss, and tracer diffusion experiments produce consistent activation energies. Dielectric loss results for Tl$^+$ and Li$^+$ beta alumina disagree with direct single crystal measurements between −100 and −50°C, as shown in Table 2. The reason for this disagreement is not known.

TABLE 2. *Activation energies observed by dielectric loss, tracer diffusion, and conductivity measurements in beta alumina*

Ion	E_a for conductivity (30)	E_a for diffusion (29)	E_a for dielectric loss (17)
Na	3.71(−180 to 820°C)	3.81(25 to 400°C)	3.70(−188 to −147°C)
K	6.67(−70 to 820°C)	5.36(200 to 400°C)	8.48(−103 to −43°C)
Ag	3.87(25 to 800°C)	4.05(25 to 400°C)	3.78(−188 to −147°C)
Rb	—	7.18(200 to 400°C)	12.10(−26 to +30°C)
Tl	8.05(−20 to 800°C)	8.22(200 to 400°C)	7.38(−103 to +67°C)
Li	8.41(180 to 800°C)	8.71(200 to 400°C)	8.65(−103 to −53°C)

All activation energies are in Kcal/mole.

The only beta″ alumina isomorph for which detailed single crystal conductivity measurements have been reported (15) is sodium beta″ alumina. Measurements were carried out on single crystals of Mg-stabilized sodium beta″ alumina using gold-blocking electrodes. Conductivity as a function of temperature was determined at 10^6 Hz. The results are shown in Fig. 6, which also includes data previously discussed for sodium beta alumina. These data indicate that sodium beta″ alumina has a larger conductivity over the temperature range shown than sodium beta alumina. More striking, however, is the curvature in the Arrhenius plot for sodium beta″ alumina. Further investigation is necessary to confirm that this observation does indeed reflect an intrinsic property of sodium beta″ alumina.

From considerations of structure it is reasonable that the conductivity of ions larger than sodium would be greater in beta″ alumina than in beta alumina. Some indication that this is true has been presented by Kummer (12) who quotes unpublished results indicating that the conductivity of potassium beta″ alumina single crystals is greater than K^+ beta alumina crystals.

Until recently, conductivity in the various protonic beta and beta″ aluminas, which include H_3O^+, H^+, and NH_4^+ beta and beta″ alumina, has been unexplored. Results have now been presented on conductivity in H_3O^+ beta and beta″ alumina that are reviewed in detail later in this chapter.

In all of the beta and beta″ aluminas it is inaccurate to discuss the specific conductivity of a compound as if it were invariant. The beta and beta″ alumina compositions are nonstoichiometric. As such, it is reasonable that their conductivities and activation energies vary with changes in stoichiometry. To date no detailed study of the relationship between conductivity and stoichiometry in these compounds has appeared.

Ion Exchange Reactions of Sodium Beta and Beta″ Alumina

One reason the beta and beta″ alumina structures have attracted great interest as solid ionic conductors is that the entire sodium ion content of each can be replaced by a variety of monovalent and divalent cations. Most ion exchange investigation has been carried out with the beta alumina structure. Yao and Kummer (29) found that Li^+, Ag^+, Tl^+, K^+, and Rb^+ exchange with Na^+ in Na^+ beta alumina exposed to $NaNO_3$–MNO_3 melts at 350 to 450°C. Partial exchange was observed for Cs^+ under similar conditions. The exchange equilibria they reported are shown in Fig. 7. Complete exchange of Na^+ by NO^+, Ga^+, and Cu^+ and partial exchange by divalent cations such as Sr^{2+}, Zn^{2+} have also been reported (27). In^+ beta alumina has been prepared by reacting Ag^+ beta alumina with molten In at 300°C for several days (29).

Several protonic beta aluminas have also been prepared. NH_4^+ beta alumina results from exchanging sodium beta alumina in molten NH_4NO_3 at 170 to 200°C (29). Saalfeld et al. (22) found that the entire sodium content of sodium beta alumina can be replaced by hydronium ions (H_3O^+) by equilibration in

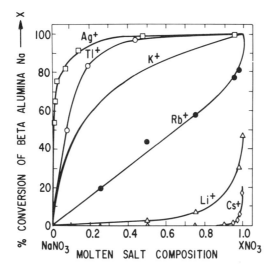

FIG. 7. Ion exchange equilibria for sodium beta alumina with various alkali nitrate melts at 350°C (29).

concentrated sulfuric acid at 295°C for several days. The exchange is accompanied by considerable gross sample decomposition. [Bettman, M., *unpublishea work* (cited in ref. 15)] reported preparing H$^+$ beta alumina by reducing Ag$^+$ beta alumina in dry hydrogen at 350°C. Silver metal is deposited on the surfaces and interstices of the crystals by the reaction.

Relatively little ion exchange chemistry of sodium beta″ alumina has been reported. Briant and Farrington (J. D. Briant and G. C. Farrington, *unpublished results*) have found that the equilibrium ion exchange curve for Li$^+$–Na$^+$ beta″ alumina in mixed LiNO$_3$–NaNO$_3$ melts at 350°C is very similar to that observed for Li$^+$–Na$^+$ beta alumina (Figs. 7 and 8). Thery and Briancon (26) found that sodium beta″ alumina readily "hydrates" in boiling water. Bettman and Peters (1) reported that a substantial proportion of the sodium content of sodium beta″ alumina is replaced by protons and an undetermined amount of water in cold HCl. Briant and Farrington (J. D. Briant and G. C. Farrington, *unpublished results*) have found that >90% of the sodium content of sodium beta″ alumina can be replaced by protons and an undetermined amount of water by equilibration (~7 days) of 1 mm-sized single crystals in sulfuric acid at 90 to 200°C. Further ion exchange reactions for beta″ alumina have not been reported, although it may reasonably be expected that the reactions found for beta alumina will also be observed for beta″ alumina.

One effect of replacing the sodium ion content of beta alumina with other monovalent cations having different ionic diameters is a change in the "width" of the conducting plane, that is, the distance between the close-packed oxygen layers above and below the conducting plane as measured between oxygen cen-

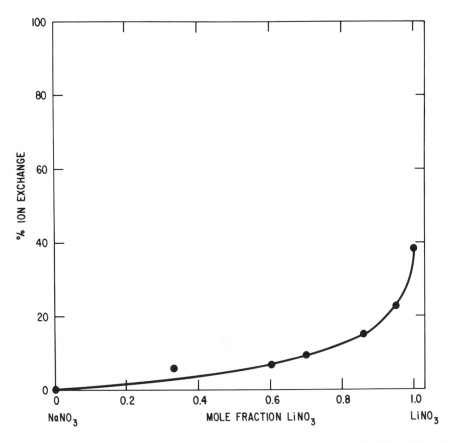

FIG. 8. Ion exchange equilibrium for sodium beta″ alumina with molten $LiNO_3/NaNO_3$ at 350°C. (J. D. Briant and G. C. Farrington, *unpublished results*).

ters. In sodium beta alumina this distance, shown in Fig. 2, is 4.76 Å. Figure 9 illustrates the variation in c_0 lattice parameter with changing ionic diameter. In general, as sodium ions are replaced by larger ions, the spacing increases. Substitution of lithium, however, produces an increase in spacing, not a decrease. It has been suggested (17) that Li^+ is too small to coordinate to both sides of the plane simultaneously and is displaced toward one close-packed oxygen layer or the other. Consequently, Li^+ ions exert less effect in drawing together the close-packed oxygen layers, and the layers spread apart slightly.

IONIC CONDUCTIVITY IN CO-IONIC AND PROTONIC BETA AND BETA″ ALUMINAS

The previous section focused on the structures and properties of alkali beta and beta″ aluminas, compositions in which a single ionic species is present.

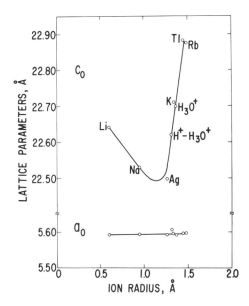

FIG. 9. Variation in c lattice parameter of beta alumina with ionic radius.

Other areas of beta alumina chemistry recently explored include (a) *protonic compositions,* in which the sodium content of beta or beta″ alumina has been replaced by protonic species such as H_3O^+, NH_4^+, or H^+; (b) *hydrated* alkali compositions, in which molecular water has diffused into the conducting planes of alkali beta or beta″ alumina; and (c) *co-ionic compositions,* in which two ionic species are present simultaneously, such as Li^+–Na^+ beta alumina. Protonic conduction in beta and beta″ alumina has been neglected until recently even though each exists in a number of protonic isomorphs. Recent measurements (J. D. Briant and G. C. Farrington, *unpublished results*[1]) indicate that H_3O^+ beta″ alumina has a high ionic conductivity, about 10^{-2} (ohm cm)$^{-1}$ at 25°C, comparable to that of sodium in the structure, making H_3O^+ beta″ alumina one of the most conductive inorganic solid proton conductors known. Recent work (G. C. Farrington, *unpublished results*) has also reported in detail on the hydration reactions of alkali beta and beta″ aluminas, relating observed behavior to ionic size and conducting plane structure. Co-ionic behavior (11,23) in Li^+–Na^+ beta alumina has also been the subject of recent investigation.

Two factors influencing ionic conductivity in beta and beta″ alumina discussed in the following sections are *ion–lattice interactions* and *ion–ion interactions.* Ion–lattice interactions are significant in affecting the mobile ion distribution in the beta and beta″ alumina conducting planes and the activation energy for ionic conductivity. A specific ion–lattice interaction has been invoked to

---

[1] General Electric Research and Development Center.

explain the increased activation energy for conduction observed in lithium beta alumina compared to sodium beta alumina in which it was suggested that lithium ions, because of their small size, bond preferentially to one of the close-packed oxygen layers bounding the conducting plane. This bond energy then contributes to the activation energy for conduction. Bond formation between mobile ions and the oxygen matrix of the conducting plane may play a dominant role in determining the properties of protonic compositions such as H^+ and H_3O^+ beta and beta" alumina, where hydrogen bonding may occur with oxygens in the conducting plane. Ion–ion interactions are less well defined compared to ion–lattice effects, but may be expected to influence activation energy and conductivity in co-ionic compositions such as Li^+–Na^+ beta alumina. Ion–ion interactions may also be central to the conduction process in H_3O^+ beta and beta" alumina.

Following are specific discussions of conductivity and chemistry in H_3O^+ beta and beta" alumina, hydration chemistry of Li^+, Na^+, and K^+ beta alumina and Na^+ beta" alumina, and conductivity and composition in co-ionic Li^+–Na^+ beta and beta" alumina.

Conductivity in H_3O^+ Beta Alumina

It is possible to replace the entire Na^+ content of sodium beta alumina with H_3O^+ ions by equilibrating single crystals of sodium beta alumina with concentrated sulfuric acid at 270 to 290°C, as Saalfeld et al. (22) first reported. Complete ion exchange of 1 mm-sized single crystals of sodium beta alumina requires about 14 days, whereas exchange of sodium ions by Li^+ at 350°C in similar size crystals occurs in several minutes. One explanation for the slow rate of exchange is slow hydronium ion diffusion within the beta alumina structure. This has since been corroborated by conductivity measurements.

The simplest hydronium beta alumina composition corresponds to complete replacement of all sodium ions by hydronium ions (Eq. 2). Saalfeld et al. (22) and Breiter et al. (5) demonstrated that at least two forms of hydronium beta alumina exist between 20 and 550°C. The low temperature composition is 3.9 $H_2O \cdot 11$ Al_2O_3, and the high temperature form is 2.4 $H_2O \cdot 11$ Al_2O_3. These two compositions reversibly interconvert with accompanying loss or gain of water between 200 to 300°C. Hydronium beta alumina irreversibly decomposes into alpha alumina and water above about 800°C (Eq. 3).

$$1.32\ Na_2O \cdot 11\ Al_2O_3 \xrightarrow[290°C]{H_2SO_4}$$

$$1.32(H_3O)_2O \cdot 11\ Al_2O_3 \equiv 3.9\ H_2O \cdot 11\ Al_2O_3 \quad [2]$$

$$3.9\ H_2O \cdot 11\ Al_2O_3 \underset{250°C}{\rightleftarrows} 2.4\ H_2O \cdot 11\ Al_2O_3 + 1.5\ H_2O \xrightarrow{>800°C}$$

$$11\ Al_2O_3 + 3.9\ H_2O \quad [3]$$

Roth et al. (19) found that finely powdered samples of hydronium beta alumina exposed to 10 torr water vapor partial pressure in nitrogen take up additional water to a composition of 4.6 $H_2O \cdot 11$ Al_2O_3 or greater. They suggested that the additional water exists as water of hydration in the beta alumina conducting plane, analogous to the hydration observed for lithium beta alumina discussed later in this chapter. Unfortunately, as noted by Roth, et al. (19) it is not possible on the basis of the data presented to rule out another explanation, that the additional water is simply adsorbed on the large surface area of the powders studied. Farrington et al. (7) found no evidence for additional water pickup beyond the composition of 3.9 $H_2O \cdot 11$ Al_2O_3 in measurements on 1 mm-sized single crystals of hydronium beta alumina.

The high temperature composition of hydronium beta alumina is 2.4 $H_2O \cdot 11$ Al_2O_3 at 350°C, as determined by Breiter et al. (5), Roth et al. (19), and Farrington et al. (7). It is invariant over water vapor partial pressures of 10 to 300 torr and decreases about 10% as the temperature is increased to 550°C. The composition corresponds to one water molecule for every two protons in the conducting plane and is a composition intermediate between H_3O^+ beta alumina and H^+–beta alumina. In this discussion, the low temperature composition is designated H_3O^+ beta alumina and the high temperature form H^+–H_3O^+ beta alumina.

Several different mechanisms may result in ionic conductivity in hydronium beta alumina, such as proton exchange among water molecules and hydronium ions, proton exchange among lattice oxygens bounding or located in the conduction plane, and hydronium ion migration. Ionic conductivity, which necessarily involves charge transfer, may also occur by a distinctly different mechanism than molecular water diffusion in the conducting plane. Although ion exchange of sodium beta alumina to H_3O^+ beta alumina is complete only after about 14 days at 290°C, partial dehydration of 1 mm-sized crystals of H_3O^+ beta alumina at 200 to 300° occurs within 1 to 2 hr. These observations suggest that the rate of neutral molecular water diffusion through the conducting planes is much faster than would be suggested from the rate of charge transfer in the structure.

Conductivity measurements on single crystals of hydronium beta alumina from 25 to 550°C were first reported by Farrington et al. (7). Conductivity was measured at DC with crystals prepared by sulfuric acid exchange of sodium beta alumina and is 7×10^{-11} (ohm cm)$^{-1}$ at 20°C. It increases with temperature from 20 to 200°C, as shown in Fig. 10, obeying an Arrhenius relationship with average σ_o and E_a of 280 (ohm cm)$^{-1}$ and 18 kcal/mole, respectively. Between 200 and 300°C hydronium beta alumina undergoes partial dehydration, which is accompanied by a decrease in conductivity. The conductivity of hydronium beta alumina above about 315°C is described by an Arrhenius plot with σ_o and E_a equal to 90 (ohm cm)$^{-1}$ and 29 kcal/mole, respectively. The conductivity decrease accompanying partial dehyration is reversible, the conductivity described by curve A in Fig. 10 being restored on cooling in a moist atmosphere to less than 200°C.

These results are consistent with a number of indirect measurements of H_3O^+ beta alumina conductivity reported in the literature. For example, Lundsgaard

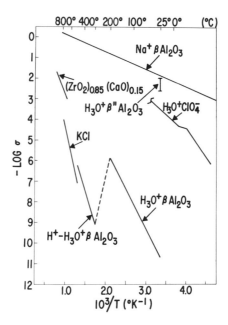

FIG. 10. Conductivity of hydronium beta and beta″ alumina as compared with sodium beta alumina, hydronium perchlorate, and calcia-stabilized zirconia.

and Brook (14) found the conductivity of a Pt–Na beta alumina–Pt cell with different concentrations of dilute HCl and H_2SO_4 in each half cell to be very small. They concluded that this stemmed from surface film formation at the interfaces, but it can also be explained by the formation of hydronium beta alumina as the result of spontaneous ion exchange at the interfaces, assuming the conductivity of hydronium beta alumina to be less than 10^{-9} (ohm cm)$^{-1}$ at 25°C. Kummer (12) found that the rate of ion exchange of sodium beta alumina with aqueous solutions of other cations to be rapidly inhibited by water. As noted by Kummer, one explanation for this behavior is that ion exchange is blocked by the formation of poorly conducting H_3O^+ beta alumina. Will (31) measured the rate of H_3O^+ replacement of Na^+ in a sample of polycrystalline Na^+ beta alumina immersed in water and found that ion exchange was accompanied by a large and progressive increase in sample resistivity. He interpreted this as the result of grain boundary water absorption, but his data can also be explained by assuming that bulk H_3O^+ beta alumina is formed in the grains of Na^+ beta alumina exposed to the water, if the bulk conductivity of H_3O^+ beta alumina is less than 10^{-9} (ohm cm)$^{-1}$ at 25°C. Farrington (6) observed that a large conductivity decrease accompanies injection of H_3O^+ into Na^+ beta alumina at a Pt–Na beta alumina interface. Although none of these reports directly measured the conductivity of H_3O^+ beta alumina, all can be explained by the finding that it has a very low conductivity.

Table 1 summarizes the preexponential factors and activation energies for

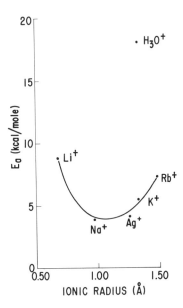

FIG. 11. Variation in activation energy for conduction in beta alumina isomorphs with ionic radius.

conductivity in various beta alumina isomorphs. Figure 11 illustrates the variation of activation energy with ionic radius. Although hydronium and potassium ions have nearly identical ionic radii, the activation energy for conductivity in hydronium beta alumina is almost four times that in potassium beta alumina. The great difference in the room temperature conductivities of hydronium beta alumina and the alkali beta aluminas arises from the difference in their activation energies. There appears to be no significant difference in preexponential factors. The results suggest that protons are tightly bound within the structure of hydronium beta alumina and not freely exchanged among hydronium ions and structural oxygen atoms. The high activation energies for conduction in H_3O^+ beta alumina and $H^+-H_3O^+$ beta alumina are probably the result of the formation of strong ion–lattice hydrogen bonds that stabilize the hydronium ions in the Beevers–Ross or midoxygen sites. The large decrease in conductivity accompanying the partial dehydration of H_3O^+ beta alumina to $H^+-H_3O^+$ beta alumina implies that the protons resulting from the dissociation and dehydration are strongly hydrogen bonded to oxygen atoms within the conducting planes. In general, the transport properties of hydronium beta alumina appear to be dominated by hydrogen bonding in the lattice.

Conductivity in H_3O^+ Beta″ Alumina

The entire sodium content of sodium beta″ alumina can be replaced by hydronium ions and an uncertain quantity of additional water by ion exchange in aqueous acids, as in the conversion of sodium beta alumina to hydronium beta alumina. But the rate of sodium ion exchange in sodium beta″ alumina crystals exposed to aqueous acids is more rapid than observed with sodium beta alumina. Thery and Briancon (26) first noted that sodium beta″ alumina crystals easily "hydrolyze" in boiling water. Bettman and Peters (1) found that a substantial exchange of sodium ions in sodium beta″ alumina crystals takes place in cold HCl. Briant and Farrington (J. D. Briant and G. C. Farrington, *unpublished results*) found that at least 80% of the sodium content of 1 mm-sized sodium beta″ alumina crystals is replaced after 3 days exposure to dilute HCl (1 M) at 90°C. Nearly complete (92 to 95%) ion exchange occurs after 3 days in concentrated H_2SO_4 at 90 to 240°C. Protons along with water molecules enter the conducting planes during the exchange process. The amount of water, which may be variable, is not known. By analogy to the chemistry of hydronium beta alumina, in which one hydrated proton or H_3O^+ ion appears to replace each sodium ion in the structure, the composition in the beta″ alumina structure is referred to here as H_3O^+ beta″ alumina. Considerably more than one water molecule for each sodium ion originally present may enter the conducting plane in H_3O^+ beta″ alumina, however.

Single crystal samples of hydronium beta″ alumina, prepared from MgO-stabilized sodium beta″ alumina by ion exchange in concentrated H_2SO_4 at 240°C, have an ionic conductivity in the range of 2×10^{-2} to 2×10^{-4} at 25°C. This is comparable to the conductivity of sodium ions in beta and beta″ alumina at 25°C. One of the few proton-containing inorganic solids exhibiting as high an ionic conductivity at 25°C is $HUO_2PO_4 \cdot 4\ H_2O$ which, although reported to have a conductivity of 5×10^{-3} (ohm cm)$^{-1}$ at 25°C (25), dehydrates around 80°C to a poorly conducting phase. Crystalline H_3OClO_4 has an ionic conductivity of 5×10^{-4} (ohm cm)$^{-1}$ at 25°C (16), but dissolves in its water of hydration at approximately 47°C (see Fig. 10). Many other salts exhibit proton conductivities at 25°C in the range of 10^{-5} to 10^{-10} (ohm cm)$^{-1}$. Glasser (10) has presented a lengthy review of this subject.

The dehydration chemistry of hydronium beta″ alumina has not been thoroughly investigated. Like hydronium beta alumina, however, it appears to undergo a composition change between 200 and 300°C in which part or all of its water content is lost. This may represent a complete dehydration of the structure to a H^+ beta″ alumina composition, whereas the analogous dehydration in hydronium beta alumina results in the loss of only about 50% of the original water content, producing a mixed H^+–H^3O^+ form. The partial dehydration of H_3O^+ beta″ alumina is accompanied by a decrease in conductivity of several orders of magnitude (see Fig. 10). A similar decrease in conductivity may result from the dehydration of hydronium beta″ alumina, although definitive measurements of conductivity above 200°C have not yet been reported.

What is most striking about the high ionic conductivity of hydronium beta″ alumina is that it presents such a contrast to the extremely low conductivity of hydronium beta alumina, which is about 10^{-11} (ohm cm)$^{-1}$ at 25°C. Yet both compounds are layer structures of similar composition, conductivity in both is two dimensional, occurring in "conducting planes" defined by close-packed oxygen layers held together by Al—O—Al bridges, and both hydronium forms show evidence of extensive hydrogen bonding on the basis of infrared spectra. But despite their structural and chemical similarities, the conductivities of the hydronium forms differ by 7 to 9 orders of magnitude.

One possible reason that hydronium beta″ alumina has such a high conductivity compared to hydronium beta alumina might be that the conduction pathways in the two structures are different, charge being transported in beta″ alumina by H^+ exchange along a continuous H_2O network in the conducting plane. The rationale behind this suggestion arises from a consideration of the geometries of the conducting planes in both structures.

As mentioned in the previous section on structure, at low temperatures in the beta alumina structure, sodium ions are found preferentially in the Beevers–Ross and midoxygen sites. Similarly, the site labeled "d" in Fig. 4 is the most favored site for sodium ion occupation in the beta″ alumina structure. The anti-Beevers–Ross sites in the beta alumina structure and the midoxygen sites in the beta″ alumina structure are the least favored for sodium ion occupation in each structure, the sites in which a diffusing sodium ion comes closest to an adjacent pair of oxygen atoms. In beta alumina one oxygen is above the anti-Beevers–Ross site in the close-packed oxygen layer and the other is below. In the beta″ alumina the two oxygens are located in the conducting plane. Considering the crystallographic radius of an oxygen ion to be 1.32 Å, the gap through which an ion must pass in the anti-Beevers–Ross site is 2.0 Å. This is smaller than the diameter of a sodium ion, 1.94 Å, but represents a "squeeze" for larger ions such as K^+ and Rb^+. In the midoxygen site of the beta″ structure there is a larger gap of 3.0 Å. As mentioned previously, this should result in higher conductivities for larger ions such as K^+ in the beta″ structure compared to the beta alumina structure, which has been reported (1).

Ionic size compared to lattice spacing does not appear, however, to explain the higher conductivity of hydronium beta″ alumina compared to hydronium beta alumina. The diameter of a potassium ion, 2.6 Å, is about the same as that of a hydronium ion, and although potassium beta alumina does have a larger activation energy for conduction than sodium beta alumina (5.5 vs 3.7 kcal/mole), it is still much smaller than the activation energy observed for hydronium beta alumina, 18 kcal/mole (7). To explain the higher activation energy of hydronium beta alumina compared to potassium beta alumina, it was suggested (7) that hydronium ions, unlike potassium ions, can hydrogen bond to oxygens surrounding the Beevers–Ross and midoxygen sites and that the energy of this bond formation contributes to the activation energy for conduction. But, infrared spectra show that extensive hydrogen bonding occurs in both hydronium beta and beta″ alumina. Were it retarding hydronium motion in one, it would be expected to do likewise in the other.

For these reasons, it is proposed here that ionic conductivity in hydronium beta" alumina is the result of rapid proton exchange among water molecules, hydronium ions, and lattice oxygen atoms in the conducting plane. According to this hypothesis, the conductivity of hydronium beta alumina is low because hydronium ions are isolated from each other. In the beta" alumina structure, each midoxygen site is filled by either an hydronium ion or a water molecule providing a continuous pathway for proton exchange and migration in the plane.

One prediction of this hypothesis is that hydronium beta" alumina crystals in the highly conductive form contain more than one water molecule for every sodium ion originally present in the structure. Another prediction is that the rate of proton motion through the structure is much more rapid than the rate of oxygen ion motion. Further experimental results are necessary to decide whether this hypothesis is correct.

Hydration Reactions of Beta and Beta" Alumina

Water undergoes two reactions with alkali beta alumina crystals: (a) ion exchange, in which M^+ beta alumina reacts at the surface to form H_3O^+ beta alumina and MOH, and (b) hydration, in which neutral water molecules migrate into the conducting planes of the crystals without displacing the alkali ion content. A diagram of the conducting planes of both beta or beta" alumina shows that there is ample space for bulk water absorption. If, for example, water molecules are placed in each unoccupied midoxygen site in sodium beta alumina, about 9% by weight of water can be accommodated by the structure. The structure of beta" alumina also has room for absorbing a considerable amount of neutral water molecules, particularly in the midoxygen sites and in sites unoccupied by sodium ions. By placing water molecules in midoxygen sites and vacant sodium sites, it is possible to fit 5 to 7 weight % of water into beta" alumina.

Absorption of molecular water into the conducting planes of alkali beta aluminas was first reported by Bettman (M. Bettman, *unpublished results*) who found that 1 mm-sized single crystals of sodium beta alumina exposed to moist air at 25°C develop broad infrared absorption in the region of 3,200 cm^{-1} that is attributable to the presence of OH groups. It was found that lithium beta alumina hydrates even more readily at 25°C and dehydrates on heating to 500°C. No evidence of hydration was found in silver, potassium, and rubidium beta alumina even after exposure to 25 torr water vapor pressure at 300°C. Evidence of water penetration of the sodium beta alumina lattice has also been found in nuclear magnetic resonance investigations. Kline et al. (11) observed that the motionally narrowed sodium resonance in powdered sodium beta alumina broadens when the sample is exposed to water vapor. A normal spectrum is restored by heating to 300°C in vacuum. The behavior is consistent with a mechanism in which water molecules penetrate the conducting planes and restrict the local sodium ion motion.

Recent investigations by Farrington (G. C. Farrington, *unpublished results*) have shown that 1 mm-sized crystals of lithium and sodium beta alumina hydrate when exposed to 10 torr water vapor partial pressure for 12 hr at 25°C. The most pronounced hydration reactions are observed with lithium beta alumina, which forms at least three distinct hydrates. These reactions are summarized in Eq. [4]. The total weight gain from composition [4-I], which is assumed to be completely dehydrated lithium beta alumina, to composition [4-IV] is 9.5%. The reactions shown in Eq. [4] appear to be reversible.

$$1.32 \, Li_2O \cdot 11 \, Al_2O_3 \cdot 6.4 \, H_2O \qquad\qquad \text{[4-IV]}$$

$$\downarrow \uparrow \quad 70°C$$

$$1.32 \, Li_2O \cdot 11 \, Al_2O_3 \cdot 2.2 \, H_2O + 4.2 \, H_2O \qquad \text{[4-III]}$$

$$\downarrow \uparrow \quad 350°C$$

$$1.32 \, Li_2O \cdot 11 \, Al_2O_3 \cdot 0.8 \, H_2O + 1.4 \, H_2O \qquad \text{[4-II]}$$

$$\downarrow \uparrow \quad 800°C$$

$$1.32 \, Li_2O \cdot 11 \, Al_2O_3 + 0.8 \, H_2O \qquad\qquad \text{[4-I]}$$

A fraction of the water molecules are bound relatively loosely in composition [4-IV] as is shown by the low temperature at which it dehydrates. It is assumed here that this loosely bound water is in the conducting planes and not in microcracks. Supporting this conclusion is the fact that if water were in cracks, similar low temperature water absorption should be observed with other alkali beta alumina crystals, since ionic substitution should have little influence on crack morphology. However, water absorption of comparable magnitude, between 25 and 100°C, is not observed with Na^+, K^+, or Rb^+ compositions.

The transition between compounds [4-II] and [4-III] occurs in a temperature range 200 to 300°C, similar to the partial dehydration reactions observed with hydronium beta and beta″ aluminas. Most intriguing is the small water loss occurring around 800°C, which must result from water tightly associated and hydrogen bonded in the structure. Its tenacity is reflected in the high temperature required for desorption.

Each of the hydrates formed has the appearance of being a definite compound, with the possible exception of composition [4-IV]. In a scan of sample weight as a function of increasing temperature (TGA), the transitions between each hydrated state occur within 100° at a scan rate of 1°C/min. From this it appears that water molecules occupy distinct and energetically well-defined coordination states within the beta alumina structure.

Hydration is also observed with sodium beta alumina and sodium beta″ alumina, but to much smaller extents than with lithium beta alumina. For example, 1 mm-sized single crystals of sodium beta alumina that have been exposed to room temperature humidity for an extended period lose about 1 weight % water between 200 to 400°C. After subsequent reexposure to 10 torr partial pressure of water vapor at 25°C for 12 hr, the crystals pick up about 1% of water, which can be desorbed around 70°C. No recurrence of the 200 to 300°C water loss is observed. Sodium beta″ alumina crystals exposed to normal room temperature humidity lose about 2% weight on heating to 200 to 300°C. No further water loss is observed to at least 850°C.

Although more data are necessary before the hydration reactions of alkali beta and beta″ alumina can be understood, it is now clear that water diffuses into the conducting planes of the structures, that a number of distinct hydrates are formed, and that in the beta alumina structure the extent of water pickup varies greatly with the alkali ion present. Lithium beta alumina undergoes the most extensive hydration reactions of the compounds discussed. On a structural basis alone, it might be expected that lithium beta alumina would be most likely to hydrate. Lithium is the smallest alkali ion, and its substitution in sodium beta alumina results in an increased spacing in the conducting planes, as the distance between the two close-packed oxygen layers expands by about 0.040 Å compared to sodium beta alumina. A structural model of beta alumina, in which water molecules with diameters of 2.6 Å are placed in each midoxygen site equidistant from the oxygen atoms that are part of the conducting plane structure, leaves space in the Beevers–Ross site for an ion of 1.43 Å in diameter. The only alkali ion smaller than this is Li^+, which has an ionic diameter of 1.36 Å. Sodium ions have larger diameters, 1.94 Å. Therefore, with this simple geometric argument alone, it is reasonable that water molecules fit into the conducting plane of lithium beta alumina more easily than in the sodium composition. Other factors such as the high polarizing power of lithium ions may also affect the relative hydration of the alkali isomorphs.

It is reasonable to expect that ionic conductivity in a hydrated alkali beta structure would be lower than in an anhydrous crystal. No definitive study of the effect of hydration on conductivity has been reported. Dunbar and Sarian (4) measured the diffusion constant for Na^+ ions in polycrystalline beta alumina before and after exposure to water vapor and found little or no difference. Unfortunately, they heated their samples to 500°C after exposure to water vapor, a treatment that would reverse any effect of hydration at a lower temperature. The observations on hydration of alkali beta alumina discussed in this section do suggest that the rate at which neutral molecular water migrates through the beta alumina structure is far more rapid than the extremely low conductivity of hydronium beta alumina would suggest. Lithium beta alumina crystals come to equilibrium with 10 torr water vapor within several hours at 25°C. The partial dehydration of hydronium beta alumina crystals occurs within 0.1 to 1.0 hr at 250°C, which is again quite rapid compared to the rate of charge transport in the structure. These observations suggest that a mechanism for rapid water diffusion throughout the beta and perhaps beta″ alumina structures exists, distinct from that involved in charge transport in the hydronium forms.

Conductivity and Composition in Co-ionic Beta Aluminas

Little work has been reported on the properties of beta and beta″ alumina in which two different monovalent cations such as Na^+–K^+, Na^+–Ag^+, and

Li^+–Na^+ are present simultaneously in the conducting plane. Yao and Kummer (29) first discussed the ion exchange equilibria between sodium beta and various alkali nitrates melts at 350°C. Their data are summarized in Fig. 7. It was recently suggested by Farrington and Roth (8) that the skewed Li^+–Na^+ ion exchange curve in Fig. 7 results from a specific interaction between Li^+ and Na^+ ions in the conducting plane of beta alumina, which imparts unusual stability to a 1:1 Li:Na composition. It was also suggested that the curious ion exchange behavior results from preferential occupation of specific nonequivalent sites within beta alumina when two or more nonidentical ions are present. The preferential occupation was also considered a manifestation of mobile ion interactions in the beta alumina conducting plane, which were termed co-ionic interactions.

If indeed the nonequivalence among the ionic sites available for mobile ion occupation in beta alumina were central in determining its ion exchange properties, then a structure such as beta″ alumina, which is chemically similar to beta alumina with the crucial exception that all mobile ion sites are identical, should display markedly different ion exchange chemistry. However, observations of Briant and Farrington (J. D. Briant and G. C. Farrington, *unpublished results*) show that the ion exchange equilibrium for Na^+ beta″ alumina in mixtures of $LiNO_3$–$NaNO_3$ at 350°C is virtually identical to that of Na^+ beta alumina (Fig. 8). There is apparently no large influence of the nonequivalent sites in the beta alumina structure compared to the equivalent sites in beta″ alumina. There is no evidence that the specific Li^+–Na^+ compositions formed in these structures are in any way unexpected from a consideration of the properties of pure sodium and pure lithium beta alumina. On the contrary, the ion exchange equilibrium between Li^+–Na^+ beta alumina and $LiNO_3$–$NaNO_3$ melts can be explained merely by assuming that the heat of formation of Li^+ beta alumina differs from that of Na^+ beta alumina or the heat of formation of $LiNO_3$ differs from that of $NaNO_3$, or both. No co-ionic interactions need be postulated to understand the equilibrium. However, it will also be shown that co-ionic effects may influence ionic conductivity and activation energy in mixed phases and that it is in ionic conductivity that co-ionic interactions should be apparent, if significant at all.

The equilibrium between Li^+–Na^+ beta alumina and $LiNO_3$–$NaNO_3$ melts is summarized in Eq. [5]. The equilibrium constant for the reaction is defined in Eq. [6] and related to the free energy change (ΔG) for the reaction by Eqs. [7] and [8]. The constant K_e is expressed in terms of the activities (a_x) of the ionic constituents within the solid beta alumina and the liquid nitrate melt. K_e may also be expressed in terms of the concentrations (C_x) of the alkali ions in the solid and the melt multiplied by appropriate activity coefficients (γ_x), as in Eq. [8].

$$Na^+\beta Al_2O_3\,(s) + LiNO_3\,(l) \underset{350°C}{\rightleftharpoons} Li^+\beta Al_2O_3\,(s) + NaNO_3\,(l) \qquad [5]$$

$$K_e = \left(\frac{{}^aLi^+}{{}^aNa^+}\right)_s \left(\frac{{}^aNa^+}{{}^aLi^+}\right)_l \qquad [6]$$

$$\Delta G_{350°C} = \Delta G_f(\text{Li}^+\beta\text{Al}_2\text{O}_3) - \Delta G_f(\text{Na}^+\beta\text{Al}_2\text{O}_3)$$
$$+ \Delta G_f(\text{NaNO}_3) - \Delta G_f(\text{LiNO}_3) \quad [7]$$

$$\Delta G_{350°C} = -RT \, ln\text{K}_e \quad [8]$$

$$K_e = \left(\frac{{}^c\text{Li}^+}{{}^c\text{Na}^+}\right)_s \left(\frac{{}^c\text{Na}^+}{{}^c\text{Li}^+}\right)_l \left(\frac{{}^\gamma\text{Li}^+}{{}^\gamma\text{Na}^+}\right)_s \left(\frac{{}^\gamma\text{Na}^+}{{}^\gamma\text{Li}^+}\right)_l \quad [9]$$

$$K_c = \left(\frac{{}^c\text{Li}^+}{{}^c\text{Na}^+}\right)_s \left(\frac{{}^c\text{Na}^+}{{}^c\text{Li}^+}\right)_l \quad [10]$$

$$-RT \, lnK_c = \overset{1}{\Delta G_f(\text{Li}^+\beta\text{Al}_2\text{O}_3 - \text{Na}^+\beta\text{Al}_2\text{O}_3)_s}$$
$$+ \overset{2}{\Delta G_f(\text{NaNO}_3 - \text{LiNO}_2)_l} + \overset{3}{RT(ln\gamma_{\text{Na}^+} - ln\gamma_{\text{Li}^+})_s}$$
$$+ \overset{4}{RT(ln\gamma_{\text{Li}^+} - ln\gamma_{\text{Na}^+})_l} \quad [11]$$

Combining Eqs. [7] to [10], K_c may be defined (Eq. [10]) in terms of two constant factors, terms 1 and 2, which are functions solely of the free energies of formation of the pure alkali beta aluminas and nitrates along with two additional terms, 3 and 4, which express nonideal effects arising when LiNO_3 and NaNO_3 melts or Li^+ beta alumina and Na^+ beta alumina phases are mixed. Terms 3 and 4, therefore, are functions of the mixed compositions in the solid and the melt. If 3 and 4 are significant compared to 1 and 2, K_c will vary with composition. If 3 and 4 are small, K_c will be essentially defined by 1 and 2.

LiNO_3–NaNO_3 mixtures are nearly ideal in behavior, that is, the free energy of formation of a specific mixed composition varies little from that predicted from a proportional averaging of the free energies of formation of the pure components. Therefore, γ_{Li} and γ_{Na} in term 4 are close to 1 for all mixtures, and term 4 is therefore nearly zero.

Term 3 reflects nonideal ionic mixing in the solid beta alumina structure. If an unusually stable composition of Li^+–Na^+ beta alumina were formed as the result of ion–ion or ion–lattice interactions in the solid, then the value of term 3 would vary with the composition of the solid, and K_c would be a function of solid composition. If, in contrast, the mixed solid behaved as an ideal mixture of the two extreme compositions Li^+ beta alumina and Na^+ beta alumina, that is, if the free energy of formation of, for example, 1:1 Li^+–Na^+ beta alumina were merely the average of the free energies of formation of Li^+ beta alumina and Na^+ beta alumina, γ_{Li} and γ_{Na} in the solid would be nearly 1 and term 3 nearly zero. K_c, then, would be essentially constant over the composition range shown in Fig. 7.

It has been observed experimentally for the equilibrium between Li^+–Na^+ beta alumina and LiNO_3–NaNO_3 melts at 350°C (29) that the value of K_c is nearly constant across the solid/liquid composition range shown in Fig. 7. That term 4 in Eq. [11] is nearly zero and that terms 1 and 2 are constants imply that term 3 must be either very small or constant over the composition range. It cannot be a strong function of composition. If specific co-ionic interactions occurred on a particular composition in the solid, the value of term 3 would vary with composition. Therefore, the simplest assumption is that term 3 is very small in comparison to 1 and 2 and that Li^+–Na^+ beta alumina compositions display nearly ideal equilibrium behavior. There appear to be no specific co-ionic interactions in the solid imparting preferential stability to a particular composition. Equilibrium between Li^+–Na^+ beta alumina and $LiNO_3$–$NaNO_3$ melts at 350°C is well described by the simple differences in the free energies of formation of the pure molten and solid phases.

The simultaneous presence of two different mobile ions within beta alumina does, however, influence ionic conductivity in the structure such that the conductivity of a mixed composition is not necessarily simply proportional to the conductivities of the extreme compositions. For example, the resistivity of 1:1 Na^+–K^+ beta alumina is not halfway between the resistivities of pure Na^+ and pure K^+ beta alumina. Radzilowski et al. (17) found that the activation energy for conduction in solid solutions of Na^+ and K^+ in beta alumina varies as in Fig. 12, so that conductivity at 25°C reaches a broad minimum around 40% K^+ replacement of Na^+. Farrington and Roth (8) observed that the conductivity of 1:1 Li^+–Na^+ beta alumina is about 5×10^{-3} (ohm cm)$^{-1}$ at 25°C compared to conductivities of Na^+ beta alumina and Li^+ beta alumina of 1.4×10^{-2} (ohm cm)$^{-1}$ and 1.3×10^{-4} (ohm cm)$^{-1}$, respectively, at the same temperature.

At least two phenomena, ion–lattice interactions and ion–ion interactions, may account for the nonlinear resistivity variation in mixed beta alumina compositions. Radzilowski et al. (17) suggested that the increase in activation energy observed as sodium ions substitute for potassium in beta alumina results from the gradual contraction of the distance between the close-packed oxygen layers defining the conducting plane. The contraction, in turn, gradually decreases the spacing through which the large potassium ion must move. The spacing in pure Na^+ beta alumina is about 0.20 Å smaller than in K^+ beta alumina. In the potassium-rich composition range, conductivity is mostly the result of potassium ion motion. The shrinkage squeezes the larger potassium ions, resulting in an increase in activation energy. In the sodium-rich region where an increasing fraction of the conductivity results from sodium ion motion, the activation energy rapidly approaches that of Na^+ beta alumina. This argument qualitatively is consistent with the experimental observations.

Although simple ion–lattice interactions may predominate in determining the conductivity of mixed beta alumina compositions, a second explanation for nonlinear conductivity variation in mixed compositions is that of specific ion–ion (co-ionic) interactions. As discussed in the section on ion conductivity in beta and beta″ alumina, one model for conductivity in beta alumina is predicated on the formation of interstitial ions pairs resulting from the excess sodium ion content of the structure. This is shown qualitatively in Fig. 5 in which interstitial pair formation is pictured for two identical ions. If the pair consists

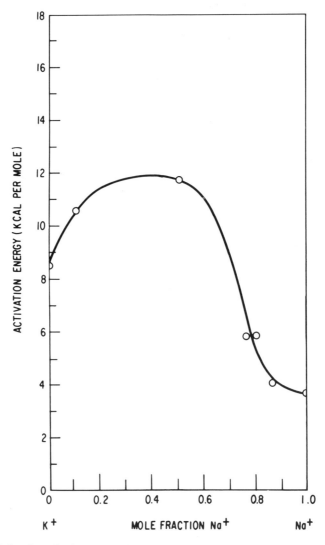

FIG. 12. Variation in activation energy from conductivity of mixed Na⁺–K⁺ beta alumina as measured by dielectric loss. (Data taken from Radzilowski et al., ref. 17.)

of two different ions, such as Li^+–Na^+, then the probability of the lithium ion migrating into the next unit cell will be different from that of the sodium ion. Ion–lattice interactions will in part determine the transition probabilities. Ion–ion interactions between the two dissimilar ions in the interstitial pair, which take into account the role of mutual repulsion in the conduction process, may also be significant. The relative magnitudes of each interaction for a particular mixed composition are not known. If, however, co-ionic interactions are major factors in determining the properties of a particular composition, as has been proposed for Li^+–Na^+ beta alumina (20), their largest influence does not appear to be in determining the composition but rather in the energetics of ionic migration and conductivity in the structure.

FUTURE RESEARCH DIRECTIONS IN BETA AND BETA″ ALUMINA

The discovery of the extraordinarily high sodium ion conductivity in the beta alumina and beta″ alumina structures stimulated the investigation of fast ion conduction in solids and the search for inorganic compounds having similar ionic conductivities. Currently, there is no question that fast ion conduction in solids, as typified by beta alumina, occurs in many structures, as shown by the variety of one-, two-, and three-dimensional ionic conductors that have been reported (24).

One of the most interesting aspects of beta and beta″ alumina chemistry is their ion exchange reactions. Perhaps the most intriguing new area of beta and beta″ alumina research is in protonic compositions in which the entire sodium ion content of the conducting planes has been exchanged with species such as H^+, H_3O^+, or NH_4^+. The chemistry of these compounds is rich. Scientifically, proton conduction in solids is a curious process because a proton is somewhat intermediate between a classic ion and an electron in its behavior. It is possible, therefore, to expect both classic proton bonding and transport mechanisms in proton conductors as well as quantum mechanical effects because of the small size of the proton. Technologically, solid proton conductors are important because a new highly conductive solid electrolyte for protons could give rise to major new devices for the storage and conversion of energy. It is reasonable to expect that investigation of transport phenomena in protonic beta and beta″ aluminas will continue. No doubt, the observation of high conductivity in H_3O^+ beta″ alumina and in $HUO_2PO_4 \cdot 4\, H_2O$ will lead to the investigation of proton mobility and conductivity in other structure types as well.

The relationship between conductivity and stoichiometry and defect chemistry in beta alumina also warrants more detailed investigation. In particular, no complete study of the variation of conductivity with composition has appeared.

Finally, a relatively unexplored area of beta alumina chemistry is that of ion motion and structure in so-called co-ionic compositions in which two different cations are present simultaneously in the conducting planes. Investigation in this area should in part illumine the nature of ion–ion interactions in the conduction process.

REFERENCES

1. Bettman, M., and Peters, C. R. (1969): The crystal structure of $Na_2O \cdot MgO \cdot 5Al_2O_3$ with reference to $Na_2O \cdot 5Al_2O_3$ and other isotypal compounds. *J. Phys. Chem.*, 73:1774–1780.
2. Beevers, C. A., and Ross, M. A. S. (1937): The crystal structure of "beta alumina" $Na_2O \cdot 11Al_2O_3$. *Z. Krist.*, 97:59–66.

3. Bragg, W. L., Gottfried, C., and West, J. (1931): The structure of β alumina. *Z. Krist.*, 77:255–274.
4. Dunbar, B. J., and Sarian, S. (1977): Effect of H_2O on Na^+ diffusivity in beta alumina. *Solid State Comm.*, 21:729–731.
5. Breiter, M. W., Farrington, G. C., Roth, W. L., and Duffy, J. L. (1977): Production of hydronium beta alumina from sodium beta alumina and characterization of conversion products. *Mat. Res. Bull.*, 12:895–906.
6. Farrington, G. C. (1976): H_2 and H_2O oxidation at a Pt/Na^+-beta alumina interface. *J. Electrochem. Soc.*, 123:833–834.
7. Farrington, G. C., Briant, J. L., Breiter, M. W., and Roth, W. L. (1978): Ionic conductivity in H_3O^+ beta alumina. *J. Solid State Chem.*, 24:311–319.
8. Farrington, G. C., and Roth, W. L. (1977): Li^+–Na^+ beta alumina—a novel Li^+ solid electrolyte. *Electrochim. Acta*, 22:767–772.
9. Frenkel, J. (1926): Über die Wärmebewegung in festen und flüssegen Kärpen (Heat motion in solids and liquids). *Z. Physik*, 35:652–669.
10. Glasser, L. (1975): Proton conduction and injection in solids. *Chem. Rev.*, 75:21–63.
11. Kline, D., Story, H. S., and Roth, W. L. (1972): Nuclear magnetic resonance diffusion studies of ^{23}Na in beta alumina, effect of water on the ^{23}Na quadruple interaction. *J. Chem. Phys.*, 57:5180–5182.
12. Kummer, J. T. (1972): β alumina electrolytes. In: *Progress in Solid State Chemistry*, Vol. 7., edited by H. Reiss and J. O. McCaldin, pp. 141–175. Pergamon Press, New York.
13. Lidiard, A. B. (1957): Ionic conductivity. In: *Handbuch der Physik*, Band 20, pp. 246–349. Springer-Verlag, Berlin.
14. Lundsgaard, J. S., and Brook, R. J. (1973): Mixed conductivity of β-alumina electrolyte in aqueous concentration cells. *J. Mat. Sci.*, 8:1519–1521.
15. Peters, C. R., Bettman, M., Moore, J. W., and Glick, M. D. (1971): Refinement of the structure of sodium β-alumina. *Acta Cryst.*, B27:1826–1834.
16. Potier, A., and Rousselet, D. (1973): Conductivité électrique et diffusion du proton dans be perchlorate d'oxonium. *J. Chim. Phys.*, 70:873–878.
17. Radzilowski, R. H., Yao, Y. F., and Kummer, J. T. (1969): Dielectric loss of beta alumina and of ion exchanged beta alumina. *J. Appl. Phys.*, 40:4716–4725.
18. Roth, W. L. (1972): Stoichiometry and structure of the superionic conductor silver beta-alumina. *J. Solid State Chem.*, 4:60–75.
19. Roth, W. L., Breiter, M. W., and Farrington, G. C. (1978): Stability and dehydration of alumina hydrates with the beta alumina structure. *J. Solid State Chem.*, 21:21–330.
20. Roth, W. L., and Farrington, G. C. (1977): Lithium-sodium beta alumina: First of a family of co-ionic conductors? *Science*, 196:1332–1334.
21. Roth, W. L., Reidinger, F., and LaPlaca, S. (1976): Studies of stabilization and transport mechanisms in beta and beta" alumina by neutron diffraction. In: *Superionic Conductors*, edited by G. D. Mahan, and W. L. Roth, pp. 223–241. Plenum Press, New York.
22. Saalfeld, H., Matthies, H., and Datta, S. K. (1968): Ein neues aluminiumoxid-hydrat β-Al_2O_3-Struktier (a new alumina hyrate with β-Al_2O_3 structure). *Ber. Dtsch. Keram. Ges.*, 45:212–215.
23. Schottky, W. (1935): The mechanism of ionic motion in solid electrolytes. *Z. Phys. Chem.*, B29:335.
24. Shannon, R. D., Taylor, B. E., English, A. D., and Berzins, T. D. (1977): New Li solid electrolytes. *Electrochim. Acta*, 22:783–796.
25. Shilton, M. G., and Howe, A. T. (1977): Rapid H^+ conductivity in hydrogen uranyl phosphate—a solid H^+ electrolyte. *Mat. Res. Bull.*, 12:701–706.
26. Thery, J., and Briancon, M. D. (1962): Sur les propriétés d'un növel aluminate de sodium $NaAl_5O_8$. *Comptes Rendus Hebdomadaires des Séances de l'Academie des Sciences*, 254:2782–2784.
27. Toropov, N. A., and Stukalova, M. M. (1939): Interchange of bases in crystals of β-Al_2O_3. Replacement of sodium in crystals of "β-alumina" with calcium, strontium and barium. *C. R. Akad. Sci. SSSR.*, 24:459–61.
28. Yamaguchi, G., and Suzuki, K. (1968): On the structures of alkali polyaluminates. *Bull. Chem. Soc. Jpn.*, 41:93–99.
29. Yao, Y. F., and Kummer, J. T. (1967): Ion exchange properties of and rates of ionic diffusion in beta-alumina. *J. Inorg. Nucl. Chem.*, 29:2453–2475.

30. Whittingham, M. S., and Huggins, R. A. (1971): Measurement of sodium ion transport in beta alumina using reversible solid electrodes. *J. Chem. Phys.,* 54:414–416.

30a. Whittingham, M. S., and Huggins, R. A. (1972): Beta alumina—prelude to a revolution in solid state chemistry. In: *Solid State Chemistry,* edited by R. S. Roth and S. J. Schneider, Jr., pp. 139–154. N.B.S. Special Publication 364, U.S. Government Printing Office, Washington, D.C.

31. Will, F. G. (1976): Effect of water on beta alumina conductivity. *J. Electrochem. Soc.,* 123:834–836.

Membrane Transport Processes, Volume 3,
edited by C. F. Stevens and R. W. Tsien.
Raven Press, New York, © 1979.

Single-File Transport: Implications for Ion and Water Movement Through Gramicidin A Channels

Alan Finkelstein and Paul A. Rosenberg

Departments of Physiology, Neuroscience, and Biophysics, Albert Einstein College of Medicine, Bronx, New York 10461

Gramicidin A forms, in lipid bilayer membranes, narrow aqueous channels through which single-file transport of ions and water may occur. The number of water molecules in such channels can be determined either from electrokinetic measurements (streaming potentials and electroosmosis) or from the magnitude of P_f/P_d—the ratio of the osmotic water permeability coefficient to the diffusive water permeability coefficient. Both methods yield a value of approximately six for the number of water molecules in a gramicidin A channel. From single-channel conductance measurements in combination with either P_d or P_f data, the water permeability of an individual gramicidin A channel can be calculated. We discuss the implications of the gramicidin A results for narrow channels in biological membranes, such as the channels induced in toad urinary bladder by antidiuretic hormone.

Single-file transport has fascinated physiologists since Hodgkin and Keynes (16) first invoked the "long pore" effect to explain why tracer potassium fluxes in giant axons fail to obey the Behn-Ussing-Teorell flux ratio equation. The possibility for construction and analysis of explicitly defined models has made this subject particularly appealing to theoreticians. Yet a certain feeling of unreality and irrelevance continues to surround it. As Heckman (14) remarks, "I don't know whether single file diffusion exists anywhere except in our imagination and on paper in the form of drawings and equations . . ." Experimentalists have not, in general, taken the theoretical consequences of single-file transport seriously; indeed, theoreticians disagree about the consequences (see below).

A case in point is single-file water transport. The prevailing view, implicitly or explicitly expressed, is that the ratio of P_f (the osmotic, or hydrodynamic, water permeability coefficient) to P_d (the tracer, or diffusional, water permeability coefficient) is equal to N (the number of water molecules in the channel[1] (5,14, 16,20,22); a minority viewpoint is that $P_f/P_d = 1$ (24,[2]26). Experimentalists

---

[1] The terms "channel" and "pore" are used interchangeably in this chapter.

[2] This paper is often cited as supporting this viewpoint, but it is not clear to us that it does.

apparently subscribe to the latter viewpoint, i.e., that P_f/P_d approaches 1 in very narrow pores (13,32). There has been to date almost no consideration of the consequences of single-file water movement on ion transport within the same channel.

In lipid bilayer membranes, gramicidin A forms very narrow channels through which single-file transport may actually occur, as suggested by their permeability to water but not to urea (8), and by molecular model building that envisions the channel as a cylinder 25 to 30 Å long with an internal radius of 2 Å (33). Thus, the theoretical possibility of single-file transport is physically realized (or approximated) and is available to experimental testing. In this chapter, we review our recent findings on ion–water interaction and water permeability in gramicidin channels, within the framework of single-file transport. We then discuss some general implications of these results and relate them to the problem of water movement through narrow pores in biological membranes—in particular, to the pores induced by antidiuretic hormone (ADH) in toad urinary bladder and in cortical collecting tubules.

THEORY

Ion–Water Interaction

Irreversible Thermodynamics

The coupling of ion and water fluxes through membranes is manifested by electrokinetic effects such as streaming potentials and electroosmosis. These phenomena are commonly treated by the formalism of irreversible thermodynamics. Although we begin with this approach, we relate the general results of that theory to the particular case of single-file transport through a permselective channel.

For a membrane (or pore) separating identical salt solutions (e.g., 0.1 M KCl) one writes, for small gradients of the driving forces, the phenomenological equations (e.g., see ref. 4):

$$I = L_{11}\Delta\Psi + L_{12}\Delta P \qquad [1a]$$

$$J_v = L_{21}\Delta\Psi + L_{22}\Delta P \qquad [1b]$$

where I is the current passing across the membrane (taken as positive when it passes from side 1 to side 2), J_v is the volume flow across it (in the same sense), ΔP is the pressure difference across the system (taken as positive when it is high on side 1), $\Delta\Psi$ is the electrical potential difference across the membrane (taken as positive when side 1 is positive), and the Ls are the "phenomenological coefficients." L_{21} and L_{12} are obtained from electroosmotic and streaming potential experiments, respectively. Thus, from Eqs. [1a] and [1b]:

$$L_{21} = \left(\frac{J_v}{I}\right)_{\Delta P = 0} L_{11} \qquad [2a]$$

$$L_{12} = -\left(\frac{\Delta \Psi}{\Delta P}\right)_{I = 0} L_{11} \qquad [2b]$$

The Onsager reciprocal relations (see ref. 4) guarantee, however, that

$$L_{12} = L_{21} \equiv L_e \qquad [3]$$

so that from Eqs. [2a] and [2b]: we have:

$$\left(\frac{J_v}{I}\right)_{\Delta P = 0} = -\left(\frac{\Delta \Psi}{\Delta P}\right)_{I = 0} \equiv \frac{\Psi_{streaming}}{\Delta P} \qquad [4]$$

Equation [4] states that the results of streaming potential experiments are predictable from the results of electroosmotic experiments, and conversely. This is a general statement, independent of pathway for ion and water movement. In most cases, $J_v \approx J_w$, the volume flow of water; for permselective pathways (e.g., gramicidin channels), this is exact (if the current-passing electrodes are reversible to the permeating ion). Multiplying and dividing Eq. [4] by F and $\overline{V}_w$, where F is the Faraday and $\overline{V}_w$ is the partial molar volume of water, we have:

$$\Psi_{streaming} = N' \frac{\overline{V}_w}{F} \Delta P \qquad [5]$$

$$N' \equiv \left(\frac{F}{I} \frac{J_v}{\overline{V}_w}\right)_{\Delta P = 0} \qquad [6]$$

N' is the number of water molecules transported per charge in an electroosmotic experiment and is directly determined from such an experiment. We see from Eq. [5], however, that it can also be determined from a streaming potential experiment. This is a general result that follows from the equivalence of streaming potentials and electroosmosis as expressed by the Onsager reciprocal relations.

Consider now the special case of a membrane with channels permselective to univalent cations, each channel containing on the average N water molecules, through which single-file transport occurs; in particular, this means that ions and water molecules cannot overtake each other. Assume further that the solution on the two sides of the membrane is so dilute that there is no more than one ion in a channel at any time. Then for every ion transported across the membrane in an electroosmotic experiment, all N water molecules in the channel must also cross. Thus, for single-file transport, N' of Eqs. [5] and [6] becomes N, the number of water molecules in a channel. That is, the number of water molecules in a channel can be determined from a streaming potential measurement.

Alternative Consideration

Equation [5], with N replacing N', can be derived for single-file transport through a univalent cation-permselective channel, from a direct consideration of streaming potential experiments, without recourse to electroosmosis and the theory of irreversible thermodynamics. Consider a membrane, containing such channels, separating identical solutions [called *(1)* and *(2)*] of a uniunivalent electrolyte (e.g., 0.1 M KCl), and let a pressure difference, ΔP, be applied across the membrane. Normally one would expect water to flow from compartment *(1)* to compartment *(2)*. Suppose, however, that each channel always contains one cation. Then in an open-circuited situation, there can be no flow, because charge would accumulate in compartment *(2)*. In other words, there is thermodynamic equilibrium, with the streaming potential producing a back electroosmotic driving force that just balances the hydrostatic pressure driving force.

For N water molecules per channel, the only kinetic unit that can cross the membrane is a cation in combination with N water molecules. Therefore, we can write for the equilibrium state:

$$N\mu_w(1) + \mu_+(1) = N\mu_w(2) + \mu_+(2) \qquad [7]$$

where μ_w is the chemical potential of water and μ_+ is the electrochemical potential of the cation. Note that neither the chemical potential of water nor the electrochemical potential of the cation is equated, but rather a particular combination of the two. The "species" that is formally in equilibrium across the membrane is the cation hydrated by N water molecules. Equation [7] can be expanded to:

$$NP(1)\overline{V}_w + P(1)\overline{V}_+ + F\psi(1) = NP(2)\overline{V}_w + P(2)\overline{V}_+ + F\psi(2) \qquad [8]$$

where ψ is the electrostatic potential of a given solution. Rewriting Eq. [8] we have:

$$\psi(2) - \psi(1) = N\frac{\overline{V}_w}{F}\Delta P + \frac{\overline{V}_+}{F}\Delta P \qquad [9]$$

The streaming potential, $\Psi_{streaming}$, is the first term on the right and is the same result (with N replacing N') obtained from irreversible thermodynamics (Eq. [5]).[3]

In deriving Eq. [9], we assumed that all channels contained one cation at all times. However, Eq. [9] remains valid if a channel never contains *more* than one ion. For although water flows through ion-free channels, the same

---

[3] $\overline{V}_+\Delta P/F$ is the contribution to the membrane potential from the difference in chemical potential of the cation. If electrodes reversible to the cation are used, this contribution is not recorded. Analogously, $J_v = J_w$ if the electrodes are reversible to the permeating ion. In practice, however, it is not necessary to use such electrodes, as it is possible experimentally to subtract the $\overline{V}_+\Delta P/F$ contribution to the streaming potential and the $\overline{V}_+J_+$ contribution to electroosmosis (29).

arguments lead to Eq. [9] for ion-containing channels and the streaming potential from these channels is not shunted by the ion-free channels.

In discussing below results obtained from gramicidin-treated lipid bilayer membranes, we refer to the equations developed in this section. Although significant hydrostatic pressure differences cannot be applied across such membranes, osmotic pressure differences, $\Delta\pi$, are readily applied with impermeant nonelectrolytes. In that case, $\Delta\pi$ replaces ΔP in the above equations. (See ref. 23 for the proper form of the phenomenological equations when $\Delta\pi$, rather than ΔP, is a driving force.)

Water–Water Interaction

We have noted above the controversy concerning the value of P_f/P_d for single-file transport. We are persuaded by the arguments that $P_f/P_d = N$, the number of water molecules in the channel, and offer a derivation in the Appendix to this chapter. Here we wish to give some intuitive feeling for the result.

The value of P_f/P_d falls continuously as the radius of the channel becomes smaller, and it corresponds remarkably well to macroscopic theory even in a pore of 4 Å radius [e.g. nystatin and amphotericin B (17)], where no more than two rows of water molecules can be accomodated. For single-file transport, however, P_f/P_d is no longer a function of radius; instead, it depends on an entirely new parameter—the number of water molecules in the channel (which is, all other things being equal, proportional to the length of the channel). The intuitive reason for this "discontinuity" is that in channels wide enough to accomodate more than a single file of water molecules, water diffuses by hopping into vacancies, more or less as in free solution. In single-file transport, however, this is precluded by definition; the only way for a water molecule to move on to the next "site" is by longitudinal, coherent motion of the entire chain. For this reason, P_d is proportional to $1/N^2$ (see Appendix), whereas P_f is proportional to $1/N$ (as one expects from macroscopic theory, where hydrodynamic flow is inversely proportional to pore length). In other words, P_f behaves more or less classically in the single-file situation, but P_d does not, because of the unique nature of the diffusion process.

RESULTS

Streaming Potentials Across Gramicidin A-Treated Membranes

In 0.01 and 0.1 M salt solutions (CsCl, KCl, and NaCl), the streaming potential is 3.0 mV per osmolal of osmotic pressure difference, with the compartment containing the nonelectrolyte osmoticant positive with respect to the opposite side (29). This means (from Eq. [5]), that 6.5 water molecules are transported per ion crossing the membrane. If we assume that single-file transport occurs through gramicidin channels and that there is rarely more than one ion in a channel, we can conclude that the gramicidin A channel contains, on the average, 6.5 water molecules.[4]

Interestingly, osmotic gradients in 0.01 M HCl produce no measurable streaming potentials. This is consistent with a Grotthus mechanism for proton transfer, as suggested (15) by the channel's "abnormally" large proton conductance. Little coupling of ion flow to water movement is expected with this conduction mechanism because proton movement proceeds through the channel down a chain of water molecules without pushing the water molecules along.

In 1 M salt solutions, the streaming potential is reduced to 2.35 mV per osmolal gradient. The smaller streaming potential means (from Eq. [5]) that fewer water molecules (5.1 instead of 6.5) are transported per ion passing through a channel. This could mean either that there are fewer water molecules in a channel or that some channels contain more than one ion (at any instant) at this higher salt concentration. The latter possibility takes on an added interest in view of recent claims (7,27,30), including some reported in this volume, of multiple ion occupancy of the gramicidin channel. The data leading to these claims—single channel conductances, I-V characteristics, and diffusion potential measurements—must ultimately be reconciled with streaming potential data. A proper analysis would yield information concerning the location of ions within the channel, since N' represents the average number of water molecules between ions in a channel with multiple ion occupancy.

Water Permeability of Gramicidin A-Treated Membranes

Since the water permeability induced by gramicidin A in lipid bilayer membranes increases linearly with conductance (i.e., with the number of channels) (28), it is convenient to normalize the permeability coefficients to some arbitrary conductance. At a conductance of 1 mho/cm² in 0.01 M NaCl:

$$P_d = 6.5 \times 10^{-3} \text{ cm/sec} \qquad [10a]$$

$$P_f = 34.2 \times 10^{-3} \text{ cm/sec} \qquad [10b]$$

(28). Thus, $P_f/P_d = 5.3$ for gramicidin channels. If we assume that single-file transport occurs in these channels, we conclude that the gramicidin A channel contains, on the average, 5.3 water molecules.

Since single-channel conductances are known [e.g., (15)], we can calculate the water permeabilities per channel, p_d and p_f, if we assume that channel conductances in densely channeled membranes (in which water permeabilities are measured) are the same as in sparsely channeled membranes (in which single-channel conductances are measured). The single-channel water permeabilities are defined by the relations:

---

[4] Levitt et al. (23) conclude, also from streaming potential data, that the gramicidin A channel contains about 12 water molecules. We have discussed elsewhere (29) what we believe is an error in their interpretation of the data that leads to this larger value.

$$p_d = \frac{P_d A}{n} \qquad [11a]$$

$$p_f = \frac{P_f A}{n} \qquad [11b]$$

where A is the membrane area and n is the number of channels corresponding to a given value of P. For the phosphatidylethanolamine membranes used in the water permeability determinations, the single-channel conductance in 0.01 M NaCl is 2.8×10^{-13} mho (based on the value of 2.8×10^{-12} mho in 0.1 M NaCl; ref. 1). With this value, we obtain from Eqs. [10] and [11]:

$$p_d = 1.82 \times 10^{-15} \text{ cm}^3/\text{sec} \qquad [12a][5]$$

$$p_f = 9.58 \times 10^{-15} \text{ cm}^3/\text{sec} \qquad [12b]$$

DISCUSSION

Number of Water Molecules in a Gramicidin A Channel

Assuming that an ion cannot overtake a water molecule within the channel, we conclude, from streaming potential measurements, that there are 6.5 water molecules per channel; assuming that water molecules cannot overtake each other within the channel, we conclude, from water permeability measurements, that there are 5.3 water molecules per channel. It is remarkable that essentially the same result is obtained from two different and completely independent sets of experiments and theories. This agreement adds further support to the belief that ion and water transport through the gramicidin channel occurs by a single-file process.

It is quite reasonable for the gramicidin A channel to contain about six water molecules. The $\pi^6_{L,D}$ helix model for the channel (33) is a cylinder 2 Å in radius and 25 to 30 Å in length which maximally accommodates only 10 water molecules (23). Interestingly, the density of water in the channel is within a factor of 2 of its density in bulk solution. Thus, each water molecule in bulk solution occupies a volume of 3×10^{-23} cc, whereas each water molecule in the channel occupies a volume of 6×10^{-23} cc if the channel contains six water molecules.

Coupling of Ion and Water Movement

The hydrodynamic permeability of a membrane, or channel, can be determined under either open-circuited or short-circuited conditions. In the former instance, the hydrostatic (or osmotic) pressure difference is opposed by a back "electrical

---

[5] This is smaller than the value reported previously (8), as discussed by Rosenberg and Finkelstein (28).

pressure" (from the streaming potential). This reduces the volume flow below that, measured under short-circuited conditions, resulting from the hydrostatic pressure difference alone. From the phenomenological Eqs. [1a] and [1b] and Eq. [3], it is easily seen that:

$$L_{22} = \left(\frac{J_v}{\Delta P}\right)_{\Delta \Psi = 0} \qquad [13a]$$

$$L_2 \equiv \left(\frac{J_v}{\Delta P}\right)_{I = 0} \qquad [13b]$$

with L_{22} and L_2 related through the equation:

$$L_2 = L_{22}(1 - Q) \qquad [14]$$

where

$$Q \equiv \frac{L_e^2}{L_{11} \cdot L_{22}} \qquad [15]$$

L_{22} and L_2 are the hydrodynamic permeability coefficients measured under short-circuited and open-circuited conditions, respectively.

Q, which can take on values between 0 and 1 (25), measures the degree of coupling between water and ion flow. When it is near 0, there is little coupling, i.e., there is negligible electroosmotic back flow in an open-circuited hydrodynamic experiment. This is expected at low salt concentrations, where at any instant most channels have no ion in them, and this indeed occurs in 0.01 M NaCl, the condition under which P_f was determined. On the other hand, if $Q = 1$, there is perfect coupling (that is, equilibrium). There is no volume flow when a ΔP (or $\Delta \pi$) is applied, because the back electroosmotic flow that would be generated by the streaming potential alone perfectly balances the flow that would be generated by the pressure gradient if the streaming potential were shorted out. (This is seen formally from Eq. [1]. When $Q = 1$, $(J_v/\Delta P)_{I = 0} = 0$ for all values of ΔP.) This could occur in our model for single-file transport through the gramicidin channel, if all channels contained the same number (> 0) of ions at high salt concentration.

For single-file transport in the gramicidin channel, or in any water-containing channel, the concept of ion mobility is somewhat unusual. By necessity, the species that moves when an ion crosses the membrane is the ion plus the N water molecules in the channel. Two situations can be envisioned. If the frictional resistance of the N water molecules is much greater than that of the ion, the mobility of the ion (i.e., ion plus N water molecules) within the channel would be the same for all ion species. This, however, would also be the mobility of a tagged water molecule, since it must also cross the membrane together with all N water molecules within the channel. Thus we have the peculiar situation that the mobility is the same for all ions and is equal to that of a water molecule. On the other hand, if the frictional resistance of the N water molecules is comparable to or much less than that of the ion, the mobility (or diffusion constant) of water would be greater in an ion-free channel than in one occupied by an ion. (In the limiting case, the ion plugs the channel and no movement, of either water or ions, occurs.)

It is possible to distinguish between these two situations by measuring p_d at both low and high salt concentrations. If the frictional contribution of the ion is significant compared to that of the N water molecules, p_d will be smaller at high salt concentrations (where most channels contain at least one ion) than at low salt concentrations (where most channels are ion free). If, however, p_d is independent of ion species and concentration, all ions have the same mobility within the channel, and observed differences in single channel conductances (15) result exclusively from differences in the partition coefficients of ions into the channel.

Single Channel Water Permeability

It is interesting to compare the water permeability of the narrow gramicidin A channel to the value predicted from a naive application of macroscopic theory. Since, as we noted earlier, the diffusion process is anomalous in single-file transport, it is appropriate to make the comparison with the hydrodynamic permeability coefficient, p_f.

For n macroscopic cylindrical pores of radius r and length L, Poiseuille's law for hydrodynamic flow is:

$$J_v \equiv \Phi_w \overline{V}_w = n \frac{\pi r^4}{8L\eta} \Delta P \qquad [16]$$

where Φ_w is the flux of water in moles per unit time and η is the viscosity (which is 0.89 centipoise for water at 25°C). If an osmotic pressure difference, $\Delta\pi (= RT\Delta c_s)$, is the driving force instead of a hydrostatic pressure difference, ΔP, Eq. [16] becomes:

$$\Phi_w = n \frac{\pi r^4 RT}{8L\eta \overline{V}_w} \Delta c_s \qquad [17]$$

where Δc_s is the concentration difference of impermeant solute producing the osmotic pressure difference. Since by definition:

$$\Phi_w \equiv n p_f \Delta c_s \qquad [18]$$

then,

$$p_f = \frac{\pi r^4 RT}{8L\eta \overline{V}_w} \qquad [19]$$

Taking $r = 2$ Å and $L = 30$ Å for the gramicidin channel, we obtain from macroscopic theory:

$$p_f = 32.3 \times 10^{-15} \text{ cc/sec} \qquad \text{(predicted from Poiseuille's law)} \qquad [20]$$

whereas experimentally:

$$p_f = 9.58 \times 10^{-15} \text{ cc/sec} \qquad \text{(experimental)} \qquad [12b]$$

It is remarkable that the hydrodynamic permeability of the very narrow grami-cidin channel, in which water molecules are probably present in only a single-file array, agrees so well (within a factor of 4) with the value predicted from a macroscopic equation (Poiseuille's law) derived from a continuum treatment of fluids. This, of course, is not the only instance of macroscopic hydrodynamic laws retaining validity at the molecular level, the most notable example, perhaps, being the success of the Stokes–Einstein equation in calculating molecular radii from diffusion constants.

Implications for Antidiuretic Hormone-Induced Channels

ADH produces striking increases in the water permeability of the luminal membrane of toad urinary bladder and mammalian collecting tubules (12,13), a result apparently brought about by the creation, or opening, of pores (9) permeable to water but not to urea (11,21). These pores are thus phenom-enologically similar to those formed by gramicidin A, and, therefore, certain inferences about them can be drawn from the results obtained on the latter.

P_f/P_d

Spuriously high values of P_f/P_d can be obtained on biological membranes because of unstirred layer problems (3). This is a recognized difficulty for P_f/P_d determinations in ADH-stimulated toad bladder (13) and cortical collecting tubules (31). Since the ADH-induced pores are tight to urea, thus implying a pore radius ≤ 2 Å, it is commonly believed that a proper determination would reveal that $P_f/P_d \approx 1$. Both the theoretical analysis of single-file transport and the results obtained on gramicidin A-treated membranes require that this expec-tation be reassessed. For if, as appears likely, single-file transport occurs through ADH-induced pores, P_f/P_d may be considerably greater than 1; in fact, a correct determination of this ratio would reveal the number of water molecules in these pores.

Number of Channels

In a maximally stimulated bladder, P_f reaches a value of $\sim 2 \times 10^{-2}$ cm/sec (2). It is most unlikely that p_f for these channels can be much greater than that for gramicidin A channels, since p_f for the latter is not much smaller than the value predicted from Poiseuille's law (see Single Channel Water Permeability, above). Using the gramicidin value (Eq. [12b]) as an approximation for p_f of the ADH-induced channels, we estimate that there are on the order of 2×10^{12} channels per cm^2 in a fully stimulated bladder. This is a useful number to bear in mind when considering and interpreting morphological correlates of ADH-induced water permeability (18,19).

Single-Channel Conductance

ADH-stimulated toad urinary bladders have conductances of approximately 10^{-3} mho/cm^2 in Ringer's solution (2). Even were this conductance attributable entirely to the ADH-induced water permeable channels (which is clearly not the case), the conductance per channel is maximally 5×10^{-16} mho. This is four orders of magnitude smaller than the conductance of a gramicidin channel in a comparable salt solution. Thus ADH-induced channels are much less ion permeable than gramicidin channels and have a much larger ratio of water permeability to ion permeability; this is also characteristic of the water permeable channels in erythrocytes (8). The physical significance of this is not apparent to us at this time; indeed, it is not clear why the conductance of gramicidin A channels is so large (10). The question deserves further study.

ACKNOWLEDGEMENT

This work was supported by grants NS 14246–01 and 5T5GM1764 from the National Institutes of Health.

The derivation of Eq. [A–19], which is closely related to Levitt's (22) derivation, was outlined to us by R. Myerson of the Institute for Advanced Study. We thank him for permission to publish it and for many fruitful discussions on single-file transport.

APPENDIX

Derivation of the Relation Between p_f and p_d for Single-File Transport

Osmosis (p_f)

Consider a pore of length L, containing N water molecules, through which single-file transport occurs (Fig. 1a). The osmotic pressure difference, $\Delta\pi$, (which is equivalent to a hydrostatic pressure difference, ΔP) between the two solutions separated by the pore is given by:

$$\Delta\pi = kT\Delta n_s \qquad [A-1]$$

where Δn_s is the difference in concentration of impermeant solute in the two solutions, expressed as molecules of solute per unit volume. (We omit the osmotic coefficient of the solute for economy.)

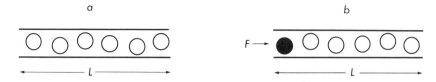

FIG. 1. Pore of length L containing water molecules through which single-file transport occurs **(a).** Pore containing a tracer water molecule where external force *(F)* acts only on the tracer molecule **(b).**

The force (F_π) exerted on the contents of the pore by the osmotic pressure difference is given by:

$$F_\pi = \frac{\rho N \Delta \pi}{L} \qquad\qquad [A–2]^6$$

where ρ is the volume per water molecule in bulk solution. The frictional drag (F_γ) on the molecules, when the contents of the pore are moving at a velocity v, can be written as:

$$F_v = N \gamma v \qquad\qquad [A–3]$$

where γ is the frictional coefficient per water molecule. In the steady state, F_π is balanced by F_γ. Equating [A–2] and [A–3] and substituting from Eq. [A–1] gives:

$$v = \frac{\rho k T \Delta n_s}{\gamma L} \qquad\qquad [A–4]$$

Using Eq. [A–4] we have for Φ_w, the flux of water:

$$\Phi_w = \frac{Nv}{L} = \frac{\rho k T N}{\gamma L^2} \Delta n_s \qquad\qquad [A–5]$$

and since for a single pore,

$$\Phi_w = p_f \Delta n_s \qquad\qquad [A–6]$$

we finally obtain:

$$p_f = \frac{\rho k T N}{\gamma L^2} \qquad\qquad [A–7]$$

Note that if we consider pores identical in all respects to that in Fig. 1 except for length, then since $N \alpha L$, we have from Eq. [A–7] that

$$p_f \; \alpha \; \frac{1}{L} \; \alpha \; \frac{1}{N} \qquad\qquad [A–8]$$

The proportionality of p_f with $1/L$ (and hence with $1/N$) is in agreement with macroscopic theory. In other words, there is nothing special about relation [A–8] for single-file transport.

---

[6] This is readily seen by considering the work, W, done when N molecules cross the pore. This is simply PV work and is given by $W = \rho N \Delta \pi$, but we also have $W = F_\pi L$. Equating these two expression for W gives Eq. [A–2].

Diffusion (p_d)

Consider a pore containing a tracer water molecule, and imagine that an external force *(F)* acts only on the tracer molecule (Fig. 1b). (This could be achieved in a centrifugal field, for example, if the density of the tracer is slightly different from that of ordinary water.) This force will cause, in the steady state, the entire contents of the pore to move with a velocity v. The applied force is balanced by the frictional drag, and we have:

$$F = N\gamma v \qquad [A-9]$$

If N^* is the mean number of tracers in the pore (which will be less than one, if we are dealing with tracer quantities), then we have for the systematic flux ($\overrightarrow{\Phi}^*$) of tracer through the pore:

$$\overrightarrow{\Phi}^* = \frac{N^* v}{L} \qquad [A-10]$$

which becomes, on substituting from Eq. [A-9]:

$$\overrightarrow{\Phi}^* = \frac{N^* F}{\gamma N L} \qquad [A-11]$$

Equation [A-11] describes the systematic flux of tracer that occurs when the force, F, acts alone. Let us now assume that we have a concentration gradient of tracer whose diffusive flux (Φ^*) balances the systematic flux. That is, we have equilibrium. [This is analogous to Einstein's (6) treatment of diffusion in free solution.] Calling n_l^* and n_r^* the concentrations of tracer in the left and right compartments, respectively, we have from the Boltzmann distribution:

$$\frac{n_l^*}{n_r^*} = e^{-FL/kT} \qquad [A-12]$$

Writing,

$$n_l^* \equiv n^* - \frac{\Delta n^*}{2}$$

$$n_r^* \equiv n^* + \frac{\Delta n^*}{2}$$

and substituting this into Eq. [A-12] we have:

$$\frac{n^* - \dfrac{\Delta n^*}{2}}{n^* + \dfrac{\Delta n^*}{2}} = e^{-FL/kT}$$

which for a small force approximates to:

$$1 - \frac{\Delta n^*}{n^*} = 1 - \frac{FL}{kT}$$

$$\Delta n^* = \frac{FL}{kT} n^* \qquad [A-13]$$

At this Δn^* there is equilibrium. That is, the diffusive flux balances the systematic flux:

$$\Phi^* = -\overrightarrow{\Phi}^* \qquad \text{[A–14]}$$

Since,

$$\Phi^* \equiv -p_d\Delta n^* \qquad \text{[A–15]}$$

we have on combining Eqs. [A–11], [A–14], and [A–15]:

$$p_d\Delta n^* = \frac{N^*F}{\gamma NL} \qquad \text{[A–16]}$$

Substituting Eq. [A–13] into this we have:

$$p_d = \frac{kT}{\gamma L^2}\frac{N^*}{Nn^*}$$

and multiplying numerator and denominator by ρ this becomes:

$$p_d = \frac{\rho kT}{\gamma L^2}\frac{N^*}{N}\frac{1}{n^*\rho} \qquad \text{[A–17]}$$

$n^*\rho$ is the fraction of molecules that are labeled in the bulk solution, and this must (for a true tracer with no isotope effect) be equal to N^*/N, the fraction of molecules that are labeled in the pore. Thus, Eq. [A–17] reduces to

$$p_d = \frac{\rho kT}{\gamma L^2} \qquad \text{[A–18]}$$

and comparing this with Eq. [A–7] we find that:

$$\frac{p_f}{p_d} = N \qquad \text{[A–19]}$$

Note from Eq. [A–18] that if we consider pores identical in all respects to that in Fig. 1 except for length, then

$$p_d\;\alpha\;\frac{1}{L^2}\;\alpha\;\frac{1}{N^2} \qquad \text{[A–20]}$$

which should be compared to [A–8]. We see that it is p_d that behaves unusually in single-file transport, it being proportional to $1/L^2$ (and hence to $1/N^2$) rather than to $1/L$, as occurs in macroscopic pores.

REFERENCES

1. Andersen, O. S. (1978): Ion transport across simple membranes. In: *Renal Function,* edited by G. H. Giebisch and E. Purcell, pp. 71–99. Independent Publishers Group, Port Washington, N.Y.
2. Andreoli, T. E., and Schafer, J. A. (1976): Mass transport across cell membranes: The effects of antidiuretic hormone on water and solute flows in epithelia. *Annu. Rev. Physiol.,* 38:451–500.
3. Dainty, J. (1963): Water relations of plant cells. *Adv. Botan. Res.,* 1:279–326.
4. de Groot, S. R. (1958): *Thermodynamics of Irreversible Processes,* pp. 185–189. North-Holland Publ., Amsterdam.
5. Dick, D. A. T. (1966): *Cell Water,* pp. 108–111. Butterworth, Washington, D.C.
6. Einstein, A. (1905): Über die von der molekularkinetischen Theorie der Wärme geforderte Bewegung von in ruhenden Flüssigkeiten suspendierten Teilchen. *Ann. Physik,* ser. 4, 17:549–

560. (For translation see his *Investigations on the Theory of the Brownian Movement,* edited by R. Fürth, pp. 1–18. Dover, New York, 1956.)

7. Eisenman, G., Sandblom, J., and Neher, E. (1978): Interactions in cation permeation through the gramicidin channel: Tl, K, Cs, Rb, Li, H, and possible anion binding. *Biophys. J.,* 22:307–340.

8. Finkelstein, A. (1974): Aqueous pores created in thin lipid membranes by the antibiotics nystatin, amphotericin B, and gramicidin A: Implications for pores in plasma membranes. In: *Drugs and Transport Processes,* edited by B. A. Callingham, pp. 241–250. Macmillan Press, London.

9. Finkelstein, A. (1976): Nature of the water permeability increase induced by antidiuretic hormone (ADH) in toad urinary bladder and related tissues. *J. Gen. Physiol.,* 68:137–143.

10. Finkelstein, A. (1977): Discussion paper. *Ann. NY Acad. Sci.,* 264:244–246.

11. Grantham, J. J., and Burg, M. B. (1966): Effect of vasopressin and cyclic AMP on permeability of isolated collecting tubules. *Am. J. Physiol.,* 211:255–259.

12. Handler, J. S., and Orloff, J. (1973): The mechanism of action of antidiuretic hormone. In: *Handbook of Physiology—Renal Physiology,* edited by S. R. Geiger, J. Orloff, and R. W. Berliner, pp. 791–814. Williams & Wilkins, Baltimore.

13. Hays, R. M. (1972): The movement of water across vasopressin-sensitive epithelia. In: *Current Topics in Membranes and Transport,* Vol. 3, edited by F. Bonner and A. Kleinzeller, pp. 339–366. Academic Press, New York.

14. Heckmann, K. (1972): Single file diffusion. In: *Passive Permeability of Cell Membranes, Vol. 3: Biomembranes,* edited by F. Kreuzer and J. F. G. Slegers, pp. 127–153. Plenum Press, New York.

15. Hladky, S. B., and Haydon, D. A. (1972): Ion transfer across lipid membranes in the presence of gramicidin A. I. Studies of the unit conductance channel. *Biochim. Biophys. Acta,* 274:294–312.

16. Hodgkin, A. L., and Keynes, R. D. (1955): The potassium permeability of a giant nerve fibre. *J. Physiol. (Lond.),* 128:61–88.

17. Holz, R., and Finkelstein, A. (1970): The water and nonelectrolyte permeability induced in thin lipid membranes by the polyene antibiotics nystatin and amphotericin B. *J. Gen. Physiol.,* 56:125–145.

18. Kachadorian, W. A., Levine, S. D., Wade, J. B., Di Scala, V. A., and Hays, R. M. (1977): Relationship of aggregated intramembranous particles to water permeability in vasopressin-treated toad urinary bladder. *J. Clin. Invest.,* 59:576–581.

19. Kachadorian, W. A., Wade, J. B., Uiterwyk, C. C., and Di Scala, V. A. (1977): Membrane structural and functional responses to vasopressin in toad bladder. *J. Membr. Biol.,* 30:381–401.

20. Lea, E. J. A. (1963): Permeation through long narrow pores. *J. Theor. Biol.,* 5:102–107.

21. Levine, S., Franki, N., and Hays, R. M. (1973): Effect of phloretin on water and solute movement in the toad bladder. *J. Clin. Invest.,* 52:1435–1442.

22. Levitt, D. G. (1974): A new theory of transport for cell membrane pores. I. General theory and application to red cell. *Biochim. Biophys. Acta,* 373:115–131.

23. Levitt, D. G., Elias, S. R., and Hautman, J. M. (1978): Number of water molecules coupled to the transport of Na^+, K^+, and H^+ via gramicidin, nonactin or valinomycin. *Biochim. Biophys. Acta.,* 512:436–451.

24. Longuet-Higgins, H. C., and Austin, G. (1966): The kinetics of osmotic transport through pores of molecular dimensions. *Biophys. J.,* 6:217–224.

25. Lorenz, P. B. (1952): The phenomenology of electro-osmosis and streaming potential. *J. Phys. Chem.,* 56:775–778.

26. Manning, G. S. (1975): The relation between osmotic flow and tracer solvent diffusion for single-file transport. *Biophys. Chem.,* 3:147–152.

27. Neher, E., Sandblom, J., and Eisenman, G. (1978): Ionic selectivity, saturation and block in gramicidin A channels: II. Saturation behavior of single channel conductances and evidence for the existence of multiple binding sites in the channel. *J. Membr. Biol.,* 40:97–116.

28. Rosenberg, P. A., and Finkelstein, A. (1978): Water permeability of gramicidin A-treated lipid bilayer membranes. *J. Gen. Physiol.* 72: 341–350.

29. Rosenberg, P. A., and Finkelstein, A. (1978): Interaction of ions and water in gramicidin A channels: Streaming potentials across lipid bilayer membranes. *J. Gen. Physiol.* 72:327–340.

30. Sandblom, J., Eisenman, G., and Neher, E. (1977): Ionic selectivity, saturation and block in

gramicidin A channels: I. Theory for the electrical properties of ion selective channels having two pairs of binding sites and multiple conducting states. *J. Membr. Biol.,* 31:383–417.

31. Schafer, J. A., and Andreoli, T. E. (1972): Cellular constraints to diffusion: The effect of antidiuretic hormone on water flows in isolated mammalian collecting tubules. *J. Clin. Invest.,* 51:1264–1278.

32. Solomon, A. K. (1968): Characterization of biological membranes by equivalent pores. *J. Gen. Physiol.,* 51:335s–364s.

33. Urry, D. W. (1972): Protein conformation of biomembranes: Optical rotation and absorption of membrane suspensions. *Biochim. Biophys. Acta,* 265:115–168.

Membrane Transport Processes, Volume 3,
edited by C. F. Stevens and R. W. Tsien.
Raven Press, New York, © 1979.

Ion Movements in Pores Formed
by Gramicidin A

S. B. Hladky, B. W. Urban, and D. A. Haydon

*Physiological Laboratory, University of Cambridge, Downing Street,
Cambridge CB2 3EG, United Kingdom*

Gramicidin A produces ion-conducting pores in thin lipid membranes. However, despite the efforts of a number of investigators, there is still much about its structure and function that is not completely understood. We present here some results obtained for the ion fluxes through the pores and discuss the model used in their interpretation. Attention to these matters has been warranted for several reasons. Gramicidin is among the simplest pore formers and has thus provided a good system in which to investigate how ions interact with a peptide-lined pore. It is also distinctly possible that the results obtained with gramicidin and the models based on them will provide useful analogies for the more complex pores that probably exist in excitable membranes.

Gramicidin A is a neutral linear peptide with both ends blocked. Although the primary structure (see Fig. 1) is known (25), the conformation in the membrane is still the subject of some controversy (see Läuger, *this vol.,* for references). By contrast, even in early 1968 there was abundant evidence that gramicidin acts on mitochondria and red blood cells by making their membranes leaky to Na and K (ref. 8) and that it induces a cation conductance in black lipid membranes (18). Shortly after, it was observed that the conductances at low levels of gramicidin were unusually noisy, which led in October 1968 to the detection of discrete conduction levels.

Subsequently, it has been possible to show that the transitions between the levels (Fig. 2) correspond to the opening and closing of pores (11,12). Thus, in Fig. 2, the first level represents a flux of 10^7 ions/second, and from this speed alone, one suspects that the conducting channel must be a pore. Fortunately, there are still more convincing data available (Table 1). If gramicidin

$$HCO-L-Val-Gly-L-Ala-D-Leu-L-Ala-D-Val-L-Val-D-Val-$$

$$-L-Tryp-D-Leu-L-Tryp-D-Leu-L-Tryp-NH-CH_2CH_2OH$$

FIG. 1. The primary structure of valine gramicidin A (25).

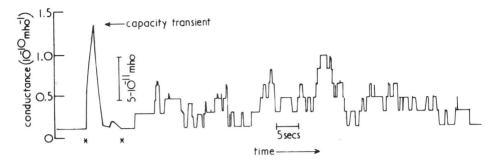

FIG. 2. A recording of the current at 100 mV across a glyceryl monooleate + *n*-hexadecane membrane containing a trace of gramicidin A. The membrane "blackened" is the time interval between the two asterisks. 0.5 M NaCl, 23°C.

is added to membranes of very different thicknesses but similar chemical nature, the current through the channel is almost the same for all of the membranes. By contrast, the mean time the channels remain open decreases substantially as the membranes become thicker—or, in other words, the channels break up less rapidly in the thinner membranes. Even more dramatically, but less reproducibly, the frequency of creation of channels increases as the membranes are made thinner. These are precisely the effects we would expect if gramicidin formed a well-defined structure that remained the same when embedded in membranes of different thicknesses. Ions passing through such a structure would see the same barriers regardless of the membrane thickness, whereas the stability of the channel and hence probably the lifetime would depend on the deformation of the membrane required to allow the channel to span it. Thus a pore of ~30 Å length might fit well into a membrane of approximately 30 Å thickness but would require considerable dimpling in order to span a thicker membrane. Further evidence that the gramicidin channels are pores has been discussed elsewhere (4,9,12,13).

The open pore is strongly selective for monovalent cations rather than anions; thus in gradients of NaCl, KCl, or Tl acetate the zero-current potentials imply transference numbers of 1 for the cation (19,27). The selectivity between cations,

TABLE 1. *Single-channel properties in membranes of various thickness*

	Hydrocarbon thickness/Å	Channel conductance/10^{-11}S	Mean channel duration/sec
gmpo + *n*-hexadecane	26	1.7	20–60
gmo + *n*-hexadecane	31	1.7	2.2
gmo + *n*-tetradecane	40	1.7	1.3
gmo + *n*-decane	47	1.7	0.4

The electrolyte was 0.5 M NaCl, temperature = 23°C. For measurement of the conductance and duration, $\Delta V = 100$ mV. The thickness was calculated from the low-field capacitance. gmo, glyceryl monooleate; gmpo, glyceryl monopalmitoleate.

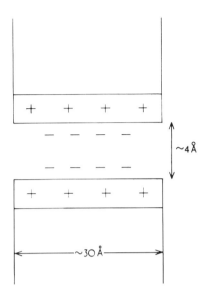

FIG. 3. Probable properties of the gramicidin pore as inferred from kinetic data and the primary structure.

as measured by conductance, can be summarized crudely by saying that if a hydrated ion is small and monovalent it goes through. The conductance sequence for the alkali cations is the same as for aqueous diffusion, complete to the point that H^+ has the highest value, which in water and ice is due to its interaction with water dipoles. The obvious inference (4) that the pore contains water was confirmed in 1973 by Finkelstein's demonstration of large fluxes of water through the pore. The demonstration of electroosmotic effects *(this vol.)* provides still more evidence.

With the qualitative flux data and the known primary structure of gramicidin A in mind, we can picture the pore schematically as a hole, probably *c.* 30 Å long and 4 Å in internal diameter (Fig. 3). The hole is lined preferentially with the negative ends of dipoles formed by the carbonyl oxygens of the peptide bonds. Neither from the structures proposed for gramicidin (see ref. 17 for figures and references) nor from the flux data is there any reason to postulate a small number of specific binding sites. The cation selectivity is conferred on the channel largely by the negative electrostatic potential attributable to the dipoles and presumably also by closer interaction of the ions with some of the carbonyl oxygens.

In order to proceed beyond this simple picture, we need to analyze the kinetic data quantitatively, which in turn requires a model. It has been clear since the first single-channel data were available that, for concentrations of NaCl and KCl above *c.* 1 M, the channel is occupied by at least one ion a large part of the time and that, for CsCl, the channel frequently contains at least two cations. This fact can be seen simply from the dependence of the channel conductance on ion activity (Fig. 4). Thus, if the channel were usually empty, the conductance would increase linearly at all activities, an occurrence that is

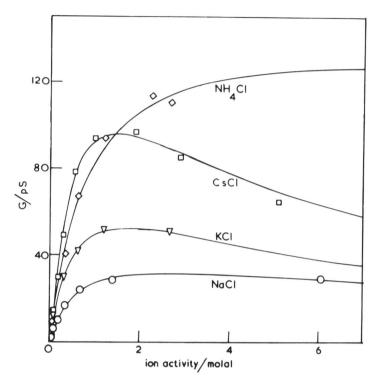

FIG. 4. The single-channel conductance versus ion activity for NaCl(○), KCl(▽), CsCl(□), and NH₄Cl(◇). 100 mV, 23°C.

not observed. Similarly, if only one ion were allowed in the channel at a time, the conductance would tend to a limiting value, corresponding to the channel's always being occupied, whereas only if a second ion could enter and block would the conductance be expected to decrease at the highest activities. Thus, as shown clearly by the data for CsCl, any model to explain the fluxes through the channel must consider either two ions in the channel simultaneously or a marked effect of the ions on the channel structure.

In general it is necessary to consider the passage of an ion through the pore either as a sequence of many short movements, as Läuger (17) has done for a pore that is at most singly occupied, or via solution of some form of continuous transport equations. However, before the equations that result from such a calculation can be used to account for data, it is always necessary to adopt simplifying assumptions to eliminate some of the constants (e.g., Läuger's assumption that all the jumps occur equally fast). By making some of the simplifying assumptions first, it is possible to avoid much of the algebra. Thus Hladky (9,10) assumed that cations in the pore spend only a small fraction of the time near the middle. This assumption is at least plausible since cations are attracted toward both the medium of higher dielectric constant (the water) and the anions, which cannot properly enter the channel. Parsegian (22) has estimated that the attraction of an ion toward the water would make the center less favorable by several kT, which would be adequate for present purposes. The advantage this assumption brings is that if the ions are indeed rarely in the middle then the pore can only exist in a relatively small number of states, e.g., empty, one ion on the left, or one ion each on the left and right. Consider a channel in a membrane separating two mixtures of two species of ion at activities a_1' and a_2' on the left and a_1'' and a_2'' on the right. If the channel is empty, an ion can enter at either end. The rate at which ions of the first species enter empty channels from the left is then (see Fig. 5):

$$A_1' a_1' X_{oo}$$

where X_{oo} is the probability a channel is empty and A_1' is a proportionality constant that depends on the applied potential, but does not vary as any of the ion activities change. The subscript specifies the species of ion moving; the prime signifies that it started on the left. If an ion of species 1 has entered from the left, then it can pass through the channel, leave to the left, or sit still while another ion enters from the opposite end. The number of each of these transitions that occur per second are $k_1' X_{10}$, $\beta_1' X_{10}$, and $D_2'' a_2'' X_{10}$, respectively. The proportionality constant for an ion to enter an occupied channel,

(a)

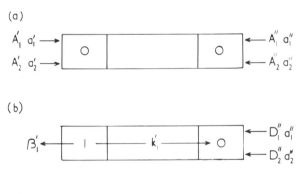

(b)

(c)

FIG. 5. The possible transitions between states in the "two-ion" model for the pore. The pore can exist in any of nine states. The proportion of the time the pore spends in each of these states is designated: X_{oo} for empty, X_{01} for an ion of species 1 on the right, X_{21} for an ion of species 1 on the right and one of species 2 on the left, etc. The number of transitions between pairs of states that occur each second are expressed in terms of the X's, the ion activities, and the proportionality constants as described in the text. Thus in (a) the rates at which ions enter empty pores are (1) for $X_{oo} \rightarrow X_{10}$, $A_1' a_1' X_{oo}$; (2) for $X_{oo} \rightarrow X_{02}$, $A_2'' a_2'' X_{oo}$; etc. The possible transitions from a single occupied state are shown in (b); those from a doubly occupied pore are shown in (c).

here D_2'', will be less than for its entry into an unoccupied channel since the ion already present will tend to repel it.[1]

If an ion of the second species does enter, then either the first or the second may subsequently leave, and the rates are $\epsilon_1' X_{12}$ and $\epsilon_2'' X_{12}$, respectively, where again we expect $\epsilon_i > \beta_i$.[1] Knock-on reactions, such as a second ion entering the side already occupied by the first, are specifically not allowed. Thus, this model is an extension to ions of part of the single-file model for neutral molecules discussed by Heckmann (5–7).

In order to describe the transport with this model, it is necessary to determine five functions of potential, $A, \beta, k, D,$ and ϵ, for each ion species present. There are, however, a number of restrictions that must be satisfied by the functions. Thus, detailed balance requires for i equal to 1 or 2:

$$\left(\frac{A_i' k_i'}{\beta_i'} \right) \bigg/ \left(\frac{A_i'' k_i''}{\beta_i''} \right) = e^{-F\Delta V/RT}$$

and

$$\frac{D_i' k_i'}{\epsilon_i'} \bigg/ \frac{D_i'' k_i''}{\epsilon_i''} = e^{-F\Delta V/RT}$$

where ΔV is the applied potential, F the Faraday, R the gas constant, and T the absolute temperature. Similarly, the independence of an equilibrium state from the route by which it is reached requires for all i and j:

$$\frac{A_1' D_2''}{\beta_1' \epsilon_2''} = \frac{A_2'' D_1'}{\beta_2'' \epsilon_1'}$$

and

$$\frac{A_2' D_1''}{\beta_2' \epsilon_1''} = \frac{A_1'' D_2'}{\beta_1'' \epsilon_2'}$$

Furthermore, partly because the precise assumptions made are not critical and partly because they are not too far from being correct, it appears permissible to assume that each A, β, etc. has the same potential dependence for each species of ion, e.g., $k_1'/k_1 = k_2'/k_2$ where k_1 and k_2 are the values at zero-applied potential.[2] Thus to fit the data for "n" ion species, adjustments can be made to the $4n + 1$ proportionality constants at $\Delta V = 0$, called hereafter rate constants, plus four constants that specify how the others depend on the applied potential.

Perhaps the simplest data to interpret are the measurements of conductance

---

[1] It has been assumed here that the repulsion between the ions depends solely on their positions and that the ion(s) present sit in the same positions regardless of species. More general equations are set out elsewhere (26).

[2] This assumption is not consistent with the variations allowed in the Ds and ϵs unless $A_i/D_i = A_j/D_j$ and $\epsilon_i/\beta_i = \epsilon_j/\beta_j$. However, errors introduced by this inconsistency are thought to be small.

at low voltage when only one species of ion is present and at the same concentration on the two sides of membrane. The model then predicts (9)

$$G = \frac{eF}{RT} \frac{A\,k\,a/\beta}{\left(1 + \dfrac{2A\,a}{\beta} + \dfrac{A\,D\,a^2}{\beta\,\epsilon}\right)\left(1 + \dfrac{2k}{D\,a + \beta}\right)}$$

where G is the conductance in the limit of small applied potentials and e is the charge on a proton. Unfortunately from conductance activity data alone it is not obvious how to assign values to the various constants. For instance, for CsCl (Fig. 6c) above 10 mM, one only needs a three-parameter polynomial to fit the data. If the point at 10 mM where the data are poorly reproducible is considered significant, then a fourth parameter is required—but still not five.

A dramatic simplification in the task of finding the constants would be possible if the concentration of ions at each end of the pore always remained at equilibrium with the adjacent solution, for then the rate of transfer would be much smaller than the other rates, and we could ignore the term

$$1 + \frac{2k}{Da + \beta}$$

As a result, there would be only three constants—the rate of transfer k and the two binding constants A/β and D/ϵ. The assumption of equilibrium, however, produces contradictions that are sufficiently basic that they remain even if we make substantial changes to the model. The simplest of these contradictions can be seen in the current-voltage relations (Fig. 6b). As ion activity is increased, currents tend to increase; in addition, the shape of the curve changes from sublinear at low activities to superlinear at high activities. In terms of the model, at low activities, the rate of entry, either Aa or Da, is slow and the flux is limited by the frequency with which ions arrive at the channel. A fast, potential-dependent step, transfer through the channel, in series with this slow weakly potential-dependent step, ion arrival, produces precisely the type of sublinear current-voltage relation seen. For high ion activity, the rates of entry are fast, and transfer is the slow step. Now the strongly potential-dependent step is rate limiting, and the current increases superlinearly with potential. If equilibrium binding is assumed, then *ad hoc* assumptions are required to explain this change in shape.

Fortunately, conductance data are not the only sort available to help us determine the constants. Myers and Haydon (19) reported on the zero-current potentials when one type of cation was on one side of the membrane and another on the other. If there are two species of ions present and the movement of any one ion is completely independent of the movements of any other, then, as is widely known, the zero-current potential, ΔV_o, can be expressed in terms of the ion activities as (23)

$$\Delta V_o = \frac{RT}{F} \ln \frac{P_1' a_1' + P_2' a_2'}{P_1' a_1'' + P_2' a_2''}$$

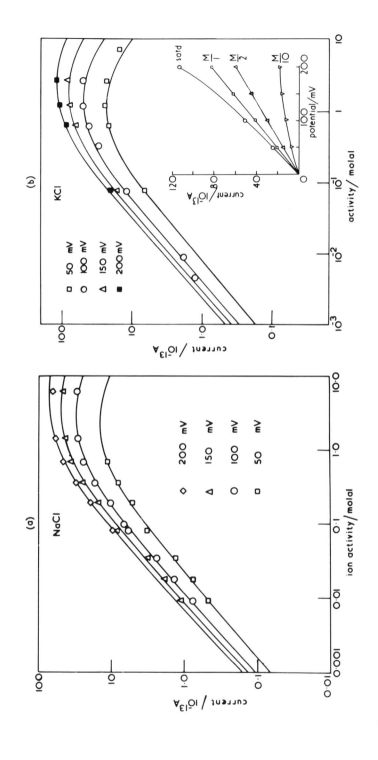

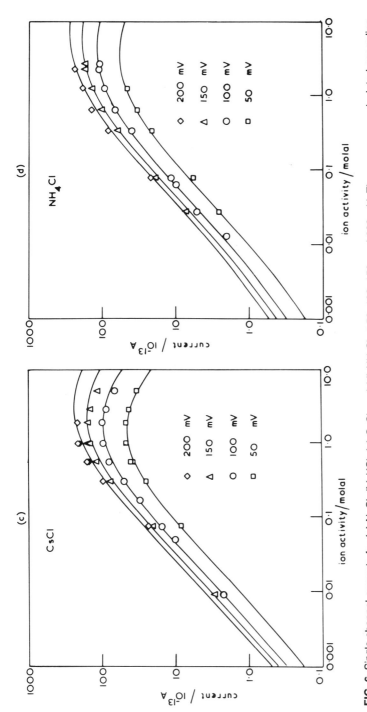

FIG. 6. Single-channel currents for **(a)** NaCl, **(b)** KCl, **(c)** CsCl, and **(d)** NH$_4$Cl at 50, 100, 150, and 200 mV. The curves are calculated according to the model using the constants in Table 2. The slopes for low activities are not equal to 1. Note the increase in separation of the curves at different potentials as the ion activities increase.

where these Ps may be regarded as generalized permeability constants because they are independent of the ion concentrations and approach the true permeabilities as ΔV approaches zero. They will frequently depend on the potential, but for certain special cases, the ratio P_2'/P_1' will not (cf. refs. 4,15). Although it is widely known that this equation applies for independent movement of the ions, it is not so widely known that this same equation, with the permeability ratio still independent of the activities, can be derived from any of the rate theories of the pore (e.g., ref. 17) that allow only one ion to enter the pore at a time. The data of Myers and Haydon (19) (see Fig. 7) clearly show that the apparent permeability ratio does depend on ion activity and thus that we are seeing effects resulting from the second ion entering the pore at much smaller activities than we had expected from the conductances. This is clear for the pair Na and K—and is even more dramatic for NH_4 and Na. There is another aspect of the data for KCl reported by Myers and Haydon that is less striking initially but even more informative on further reflection: at as low as 10 mM,

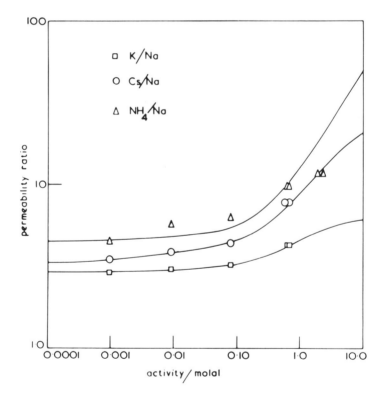

FIG. 7. Apparent permeability ratios versus activity calculated from the zero-current potentials. Na is present on one side of the membrane, and the indicated ion is present at the same activity on the other. The separation of the double symbols indicate the uncertainties in the activity coefficients. The curves are calculated from the model using the constants in Table 2.

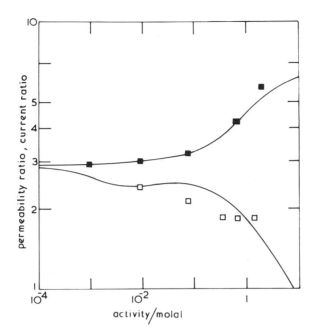

FIG. 8. Comparison of the apparent permeability ratios, ■, and conductance ratios at 100 mV, □, for KCl and NaCl. The curves are calculated from the model using the constants in Table 2. Using these constants, the model predicts, and experiments at 0.1 M have shown, that the conductance ratios are nearly independent of the applied potential.

the observed permeability ratio is not equal to the ratio of the conductances for the two species of ions (see Fig. 8). We have been reluctant to put much confidence in the absolute values of the conductances for low activities—but we have also reasoned that any effects distorting the absolute values should largely cancel in the ratios and thus that this discrepancy between permeability and conductance ratios cannot be ignored. The discrepancies for NH_4 and Cs are larger. Since independent movement of ions requires that the permeability and conductance ratios measured at the same potential be equal, the failure of this equality implies that at 10 mM the channels are already occupied by an ion a large portion of the time.[3]

The model predicts that the conductance can be proportional to activity in two different modes—independence, where $Aa/\beta \ll 1$, and singly occupied channels, where $Aa/\beta \gg 1$, $\beta \ll Da \ll k$, and $Da/\epsilon \ll 1$. In this second mode, ions leave the channel only just after a second ion enters. Thus the rate-limiting step is entry of the second ion, not transfer in the singly occupied pore or exist from the doubly occupied pore. The experimental finding that the conductance is nearly proportional to a in the region of 100 mM, where from the

---

[3] See Urban et al, ref. 27, for another attempt to fit the data.

zero-current potentials the pore is already occupied by one ion and the effects of a second are being felt, thus implies $\beta \ll k$ at least for K, Cs, and NH$_4$. Conductance data above 100 mM thus, in effect, allow the determination of the three constants D, k, and ϵ using

$$G = \frac{eF}{RT} \frac{Dak}{\left(2 + \dfrac{Da}{\epsilon}\right)(Da + k)}$$

The two remaining constants can then be determined from the conductances at low activities and comparison of the model predictions (9,26) with experimental values for the zero-current potential (19,26). In practice, since $\beta \ll$ k is not always valid, the curves are all fitted at once using a computer curve-fitting routine with an informed initial guess (26). The closeness of the fit can be seen in Figs. 6, 7, and 8. The rate constants used are described in Table 2.

As Neher reports *(this vol.)*, the thallous ion produces the most blatant deviations from simple behavior in the fluxes through gramicidin channels. The effects it produces thus require some discussion. Over the concentration range from 1 to 200 mM, Tl conductances are similar to those attributable to K (1,21,26). Thus, in any intuitive sense the permeability of the channel to Tl is of the order of twice that to K. However, by calculating permeability ratios from zero-current potentials, Eisenman and co-workers (see ref. 1) have found P_{Tl}/P_K as large as 50. The explanation of apparent values of P_{Tl}/P_K much greater than 2 must, of course, be interaction of the ions in the pore. In the model described here, high values of the apparent permeability ratio are obtained under precisely those conditions where the pore is usually occupied by a Tl ion. Any K that then enters promptly jumps back out on the same side—and thus the current, and hence the zero-current potential, is determined largely by the distribution of Tl. In other words, it is not that Tl is particularly permeable, but rather that the Tl-occupied pore is effectively impermeable to K.

Neher (20) has demonstrated this effect directly using mixtures of 2 M Na acetate and 1 M Tl acetate (E. Neher, *personal communication*). Even though the conductance with 1 M Tl substantially exceeded that with 2 M Na, the conductance of a mixture of 98% 2 M Na and 2% 1 M Tl was substantially lower than either. Thus, a small concentration of Tl (20 mM), which would, by itself, produce a much smaller conductance, blocks the conductance due to

TABLE 2. *Fitted rate constants*

	NaCl	KCl	CsCl	NH$_4$Cl
$A/10^8 \text{M}^{-1}\text{sec}^{-1}$	0.55	1.6	1.8	2.4
$\beta/10^5 \text{sec}^{-1}$	4.5	3.9	2.9	2.1
$k/10^8 \text{sec}^{-1}$	0.13	0.27	0.82	6.8
$D/10^8 \text{M}^{-1}\text{sec}^{-1}$	0.53	1.4	1.6	1.2
$\epsilon/10^8 \text{sec}^{-1}$	2.6	2.1	1.6	0.63

a large amount of Na (1.96 M). Just such blocking effects are predicted by the model described here. The essential features are that the pore must be singly occupied most of the time and that the rate constant of exit, ϵ, from the doubly occupied pore must be substantially less for the blocker than for the blocked ion. A fictitious example that displays blocking is given in Fig. 9.

In this chapter the emphasis has been to provide qualitative explanations for the kinetic effects observed with gramicidin. Much closer fits to the data will inevitably be possible if D and ϵ are allowed to depend on both species of ions present (26), since then the data for each pair of ions can be fitted almost without regard for the data for any other ions. Thus, it is difficult to conceive

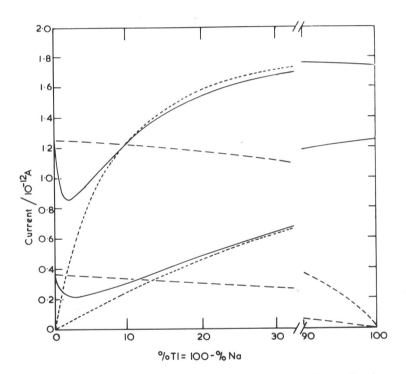

$\%$Tl = 100 - $\%$ Na

FIG. 9. A theoretically predicted blocking effect. The solid curves are the predicted currents at 50 mV for the indicated mixtures of Tl acetate and Na acetate at total concentrations of 0.1 M *(lower)* or 1 M *(upper)*. The dotted curves are the currents for Tl alone present at the same concentration as in the mixture; the dashed curves are for Na alone. For greater than 90% Tl, the predicted currents in the mixture and in Tl alone superimpose. The constants that have been arbitrarily assumed to apply are:

$$A_{Tl} = 5.4 \cdot 10^8 M^{-1} sec^{-1}, \ \beta_{Tl} = 6.1 \cdot 10^4 sec^{-1}, \ k_{Tl} = 6.7 \cdot 10^8 sec^{-1},$$
$$D_{Tl} = 3.7 \cdot 10^8 M^{-1} sec^{-1}, \ \epsilon_{Tl} = 1.5 \cdot 10^7, \ A_{Na} = 7.7 \cdot 10^7 M^{-1} sec^{-1},$$
$$\beta_{Na} = 1.1 \cdot 10^6 sec^{-1}, \ k_{Na} = 1.3 \cdot 10^7, \ D_{Na} = 5.3 \cdot 10^7, \ \epsilon_{Na} = 2.6 \cdot 10^8.$$

Since the predicted currents in the mixtures are almost unaffected by the change, A_{Na} and β_{Na} have been altered from the values in Table 2 in order to satisfy $A_{Na}/D_{Na} = A_{Tl}/D_{Tl}$ as well as the requirement that $(A_{Na}/\beta_{Na})/(D_{Na}/\epsilon_{Na}) = (A_{Tl}/\beta_{Tl})/(D_{Tl}/\epsilon_{Tl})$.

of any further critical tests of the model that can be based on current measurements alone.

There is, however, a test of the model that should be possible using tracers. From either the independence principle (14,28) or models allowing only one ion in the pore at a time (17), the ratio of the unidirectional (tracer) fluxes across the membrane

$$J{\rightarrow} \text{ and } J{\leftarrow}$$

will be related to the potential by

$$\frac{J{\rightarrow}}{J{\leftarrow}} = e^{-F\Delta V/RT}$$

Single filing is normally said to occur whenever it is observed instead that

$$\frac{J{\rightarrow}}{J{\leftarrow}} = e^{-nF\Delta V/RT} \text{ for } n > 1$$

As Hodgkin and Keynes (16) noted, this relation in turn implies that the conductance and the unidirectional flux $(J = J{\rightarrow} = {\leftarrow}J)$ at $\Delta V = 0$ are related by

$$\frac{G}{J} = \frac{neF}{RT}$$

Using this equation, the present model predicts

$$n = \frac{(Da + \beta)(4k + 2\beta + Da)}{(Da + 2\beta)(2k + \beta + Da)}$$

Thus for $Da \ll \beta$, $n = 1$ as it must, whereas for $\beta \ll Da \ll k$, $n = 2$, which is the maximum value allowed by the model. The rate constants reported here for K^+ predict $n \geq 1.7$ for activities between 2.0 and 100 mM, whereas for Cs and NH_4, $n > 1.8$ between 20 and 200 mM.

In summary, there is now substantial evidence from conductances, permeability ratios, and blocking effects that gramicidin channels are indeed occupied by one or more ions at surprisingly low concentrations. These effects have been described using a conceptually simple model that is adequate to describe the data. Of course, more complicated models, such as that of Sandblom et al. (24), that have more ions interacting with the channel and yet more adjustable constants can also be made to fit the data. However, we see neither a structural basis nor any requirement for the simultaneous presence of more than two cations in the pore.

REFERENCES

1. Eisenman, G., Sandblom, J., and Neher, J. (1977): Ionic selectivity, saturation, binding, and block in the gramicidin A channel: A preliminary report. In: *Metal-Ligand Interactions in Organic Chemistry and Biochemistry,* Part 2, edited by B. Pullman, and N. Goldblum. Reidel, Dordrecht-Holland.

2. Finkelstein, A. (1973): Aqueous pores created in thin lipid membranes by the antibiotics nystatin, amphotericin B and gramicidin A: Implications for pores in plasma membranes. In: *Drugs and Transport Processes,* edited by B. A. Callingham. Macmillan Press, London.
3. Goldman, D. E. (1944): Potential, impedance and rectification in membranes. *J. Gen. Physiol.,* 27:37–60.
4. Haydon, D. A., and Hladky, S. B. (1972): Ion transport across thin lipid membranes: A critical discussion of mechanisms in selected systems. *Q. Rev. Biophys.,* 5:187–282.
5. Heckmann, K. (1965): Zur Theorie der "single file"—Diffusion I. *Z. Physik. Chem. N.F.,* 44:184–203.
6. Heckmann, K. (1965): Zur Theorie der "single file"—Diffusion, II. *Z. Physik. Chem. N.F.,* 46:1–25.
7. Heckmann, K. (1968): Zur Theorie der "single file"—Diffusion, III. *Z. Physik. Chem. N.F.,* 58:206–219.
8. Henderson, P. J. F., McGivan, J. D., and Chappell, J. B. (1969): The action of certain antibiotics on mitochondrial erythrocyte and artificial phospholipid membranes. *Biochem. J.,* 111:521–535.
9. Hladky, S. B. (1972): The Mechanism of Ion Conduction in Thin Lipid Membranes Containing Gramicidin A. Ph.D. Thesis, University of Cambridge.
10. Hladky, S. B. (1973): Pore or carrier? Gramicidin A as a simple pore. In: *Drugs and Transport Processes,* edited by B. A. Callingham, pp. 193–210. Macmillan Press, London.
11. Hladky, S. B., and Haydon, D. A. (1970): Discreteness of conductance change in bimolecular lipid membranes in the presence of certain antibiotics. *Nature,* 225:451–453.
12. Hladky, S. B., and Haydon, D. A. (1972): Ion transfer across lipid membranes in the presence of gramicidin A. *Biochim. Biophys. Acta,* 274:294–312.
13. Hladky, S. B., Gordon, L. G. M., and Haydon, D. A. (1974): Molecular mechanisms of ion transport in lipid membranes. *Ann. Rev. Phys. Chem.,* 25:11–38.
14. Hodgkin, A. L., and Huxley, A. F. (1952): Currents carried by sodium and potassium ions through the membrane of the giant axon of Loligo. *J. Physiol.,* 116:449–472.
15. Hodgkin, A. L., and Katz, B. (1949): The effect of Na$^+$ on the electrical activity of the giant axon of the squid. *J. Physiol.,* 108:37–77.
16. Hodgkin, A. L., and Keynes, R. D. (1955): The potassium permeability of a giant nerve fibre. *J. Physiol.,* 128:61–88.
17. Läuger, P. (1973): Ion transport through pores: A rate theory analysis. *Biochim. Biophys. Acta,* 311:423–441.
18. Mueller, P., and Rudin, D. O. (1967): Development of K$^+$-Na$^+$ discrimination in experimental bimolecular lipid membranes by macrocyclic antibiotics. *Biochem. Biophys. Res. Commun.,* 26:398–404.
19. Myers, V. B., and Haydon, D. A. (1972): Ion transfer across lipid membranes in the presence of gramicidin A. II. The ion selectivity. *Biochim. Biophys. Acta,* 274:313–322.
20. Neher, E. (1975): Ionic specificity of the gramicidin channel and the thallous ion. *Biochim. Biophys. Acta,* 401:540–544.
21. Neher, E., Sandblom, J., and Eisenman, G. (1978): Ionic selectivity, saturation and block in gramicidin A channels. II. Saturation, behaviour of single channel conductances and evidence for the existence of multiple binding sites in the channel. *(Submitted for publication.)*
22. Parsegian, A. (1969): Energy of an ion crossing a low dielectric membrane—solutions to four relevant electrostatic problems. *Nature,* 221:884–846.
23. Patlak, C. S. (1960): Derivation of an equation for the diffusion potential. *Nature,* 188:944–945.
24. Sandblom, J., Eisenman, G., and Neher, E. (1977): Ionic selectivity saturation and block in gramicidin A channels. I. Theory for the electrical properties of ion selective channels having two pairs of binding sites and multiple conductance states. *J. Membr. Biol.,* 31:383–417.
25. Sarges, R., and Witkop, B. (1965): Gramicidin A. V. The structure of valine- and isoleucine-gramicidin A. *J. Am. Chem. Soc.,* 87:2011–2020.
26. Urban, B. W. (1978): The kinetics of ion movement in the gramicidin channel. Ph.D. Thesis, University of Cambridge.
27. Urban, B. W., Hladky, S. B., and Haydon, D. A. (1978): The kinetics of ion movements in the gramicidin channel. *Fed. Proc. (In press.)*
28. Ussing, H. H. (1949): The distinction by means of tracers between active transport and diffusion. *Acta Physiol. Scand.,* 19:43–56.

Membrane Transport Processes, Volume 3,
edited by C. F. Stevens and R. W. Tsien.
Raven Press, New York, © 1979.

Ionic Interactions in Potassium Channels

Ted Begenisich

Department of Physiology, University of Rochester School of Medicine and Dentistry,
Rochester, New York 14620

The movement of ions through the channels in excitable membranes shares many properties of ionic movement in aqueous solutions. Consequently, ionic current flow through membranes has often been treated like aqueous diffusion. The fundamental assumption in such treatments of ionic fluxes is that the ions move independently of one another. As formulated by Hodgkin and Huxley (11), ionic fluxes are independent if the chance that any individual ion will cross the membrane in a specified interval of time is independent of the other ions present. This definition means that the movement of any ion of type X must be independent of the presence of any other ions of type X and of all other types of ions. Obvious ways in which systems do not have independence are with ion single filing or saturating ion-binding sites.

In 1955 Hodgkin and Keynes (12), using long-lasting depolarizations, showed nonindependent movement of K ions across the membrane of *Sepia* axons. Since the Na channel responsible for initiation of the action potential inactivates very rapidly on depolarization, potassium ions were probably not moving through this pathway—even though this channel is permeable to K as well as to Na ions. But since the K channels (partly responsible for termination of the action potential) also inactivate on prolonged depolarization (7), the K efflux measured by Hodgkin and Keynes was presumably mainly through inactivated K channels and the so-called leakage pathway. Hodgkin and Keynes analyzed their data in terms of a single-filing mechanism. They suggested the presence of two to three binding sites in the K channel and that ions moved without passing each other by jumping from site to site across the membrane.

With the introduction of internal perfusion techniques in 1962 (3), it became possible to measure tracer influx and efflux under voltage clamp conditions. Such techniques were applied by Bezanilla et al. (6). They found that the Na[22] influx was not reduced by changing the internal solution from 550 mM KF to 450 mM KF + 100 mM NaF. In fact, the influx was somewhat increased. It may be that internal K ions can impede Na influx, and what they observed was a reduction in this inhibition when they reduced the K concentration. Hille (8) has considered this type of behavior and has suggested that there is a binding site for cations in the Na channel of myelinated nerve with a dissociation constant of 220 mM for K ions and 368 mM for Na ions. Begenisich and

Cahalan (4) have described a similar binding site in the Na channel of squid giant axons with equilibrium constants (at zero membrane potential) of about 900 and 420 mM for internal Na and K, respectively. The increase in influx found by Bezanilla et al. (6) is consistent with these results.

The measurements of Hille (8) and Begenisich and Cahalan (4) were all on electrical currents, not unidirectional fluxes. Certainly such information is very valuable in determining whether or not ion movement is independent. But, as Läuger (13), Hille (9), Armstrong (1), and others have pointed out, it is possible for electric currents to obey independence even with ionic interactions.

EXPERIMENTS

Recently, we (5) measured ^{42}K efflux from internally perfused squid giant axons under voltage clamp conditions. In these experiments, the axons were internally perfused with a solution containing 400 mM potassium labeled with ^{42}K. The artificial seawater (ASW) bathing the axons contained tetrodotoxin to block Na channels and various K concentrations up to 75 mM. The holding potentials for all axons was set to −78 mV, and the fluxes were measured with 20, 25, 30, or 40 mV depolarizations lasting 20 or 30 msec. These pulses opened a sufficient number of K channels to produce measurable flux ratios, yet were small enough to minimize the accumulation of K ions in the periaxonal space (2). Even so, some K loading into the 5 and 20 mM K–ASW solutions was detected from the potassium current reversal potential at the end of the depolarizing pulse. We, therefore, measured the size of the periaxonal space and found an average value of about 370 Å (range 293 to 480 Å, six axons), in excellent agreement with the value of 357 Å found by Adelman et al. (2). With our measured values of the space size and using the value for the permeability of the space from Adelman et al. (2), we could calculate the time-dependent and time-average concentration of K ions just outside the membrane. However, with our small and short depolarizations, this more complicated and tedious procedure did not give average $[K]_o$ values significantly different from those obtained by simply averaging the nominal ASW–K concentration and the value computed from the reversal potential. We, therefore, used the simpler method, and the values of $[K]_o$ in Fig. 3 are these "average" values.

Figure 1 shows ^{42}K efflux into ASW containing first 20 and then 5 mM K. After approximately 7½ min with the membrane potential held at −78 mV, the voltage was clamped once every 300 msec to a potential of −53 mV for 30 msec. The isotope efflux quickly increased during this 7½-min period to a constant value and returned to the "resting" value on cessation of the pulses. The ASW solution was then replaced with one containing 5 mM K. After another 7½-min control period, the same pulse pattern was repeated. The efflux during repetitive depolarization was now about twice that in the 20 mM solution. Any

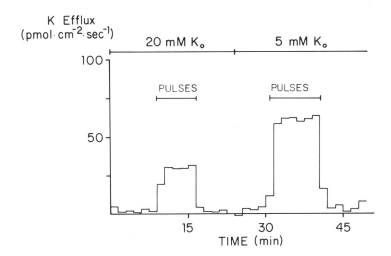

FIG. 1. Potassium efflux from an axon internally perfused with a solution containing 400 mM K$^+$ (labeled with ^{42}K) into ASW containing 20 or 5 mM K$^+$. The perfusion solution contained (in mM): K glutamate, 320; KF, 50; sucrose, 260; K phosphate, 30; pH 7.3. The axon membrane voltage was held at −78 mV throughout the experiment, except during the periods indicated by the horizontal bars when the axon was depolarized by a 25-mV, 30-msec rectangular pulse every 300 msec.

loss of K channels from axon deterioration would cause an underestimation of this difference. This result is clearly not consistent with independent ion movement.

A change in the efflux as observed in Fig. 1 could result if [K]$_o$ affected the gating properties of the channel. In particular the efflux in the high solutions could be decreased if high [K]$_o$ slowed the opening of channels or reduced the number of open channels. Our electrical measurements showed that the maximum K conductance was about the same in high and low K and the relationship between K conductance and membrane potential was only slightly altered. The effect of [K]$_o$ on potassium channel opening times are shown in Table 1. There is no clear trend. The large standard errors in Table 1 reflect

TABLE 1. *Potassium channel kinetics and potassium concentration—$t_{1/2}$ and [K]$_o$*

	[K]$_o$		
V_m	5 mM	20 mM	50 mM
(mv)	(msec)	(msec)	(msec)
−38	5.1	5.9 ± 1.35(3×)	6.4 ± 2.7(2×)
−48	7.0 ± 1.4(5×)	7.9 ± 2.2(7×)	—
−53	8.4	6.8	—
−58	10.6 ± 1.3(2×)	8.1 ± 2.4(3×)	6.0

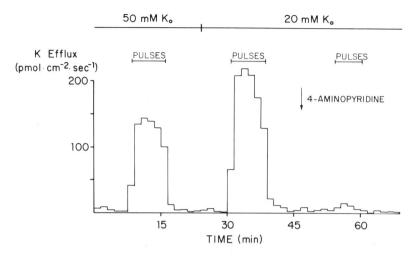

FIG. 2. Effect of external potassium ions and 4-aminopyridine on ^{42}K efflux. For experimental procedures, see Fig. 1. Holding potential was −78 mV throughout the experiment. During the periods indicated, the axon was depolarized by a 40-mV, 30-msec rectangular pulse every 200 msec.

axon-to-axon variations in time constants as well as variations within a single axon. Repeated measurements of $t_{1/2}$ values on the same axon at the same $[K]_o$ concentration showed an approximate 15% variation. Therefore, the reduction of K efflux by high $[K]_o$ can be ascribed to a direct effect of external K ions on outward moving K ions and not to indirect effects on the gating process.

Another experiment is shown in Fig. 2 for another axon in 50 and 20 mM K–ASW. Here, the membrane potential was depolarized once every 200 msec to −38 mV for 30 msec. Many more K channels are open at this potential than at −53 mV, so the efflux here is much larger than in Fig. 1. Again, the efflux is lower in the high K solutions. After about 45 min, 2 mM 4-aminopyridine was added to the ASW. This compound has been shown to be a rather specific blocker of potassium channels in these axons (14). In this experiment, the potassium current was reduced to less than 10% of control; the extra ^{42}K efflux was also reduced to very low levels by this treatment. This shows that a very large part of the measured efflux is through potassium and not "leakage" channels.

THEORY

Hodgkin and Keynes (12), Hladky and Harris (10), Armstrong (1), and others have described a simple model for K channels that can be called a "knock-on" model. In this model, there is a site within the K channel that is always filled. When a particular ion occupies the site, it leaves only by being "knocked" off by another ion, which then occupies the site. This can formally be written down as follows:

$$K_o + KS \underset{\beta}{\overset{\alpha}{\rightleftarrows}} KS + K_i$$

$$K_o^* + KS \underset{\beta}{\overset{\alpha}{\rightleftarrows}} K^*S + K_i$$

$$K_o + K^*S \underset{\beta}{\overset{\alpha}{\rightleftarrows}} KS + K_i^*$$

$$K_o^* + K^*S \underset{\beta}{\overset{\alpha}{\rightleftarrows}} K^*S + K_i^*$$

where radioactive ions are labeled by an asterisk. K_o and K_i represent external and internal K ions, respectively. KS and K^*S denote site occupancy by an unlabeled and a radioactive K ion, respectively. For the case where ^{42}K is only inside the axon ($K_o^* = 0$, $K_i = 0$) and under the assumption of steady state and a fixed number of sites, the following relationships can be obtained:

$$\text{Flux ratio: } F_{in}/F_{eff} = \left(\frac{K_o}{K_i^*}\frac{\alpha}{\beta}\right)^2 \quad [1]$$

$$\text{Efflux/site: } F_{eff} = F_{eff_{max}}\left(\frac{K_m}{K_m + K_o}\right) \quad [2]$$

where

$$K_m = K_i^* \beta/\alpha$$
$$F_{eff_{max}} = K_i^* \beta$$

The ratio of forward and reverse rate constants α/β is $\exp(-V_m F/RT)$, independent of the location of the site. In this equation, V_m is the membrane voltage; F, R, and T have their usual meanings, and RT/F is approximately 25 mV at 10° C. This can be derived from Eyring rate theory; however, it should be obvious that this must be so to correctly predict the equilibrium or Nernst potential.

Figure 3 shows a summary of our data for test depolarizations to -38 and -48 mV. The K concentration for one-half inhibition (K_m) values for -38, -48, and -58 mV were found to be 31.9 ± 6.3 (7×), 8.6 ± 1.9 (12×), and 3.1 ± 0.7 (6×) mM, respectively: a much steeper voltage dependence than predicted by the ratio α/β. This may reflect uncertainty in the data or may suggest the presence of more than one site in the membrane. Further work is necessary to determine just how many sites there are. This simple model can be made

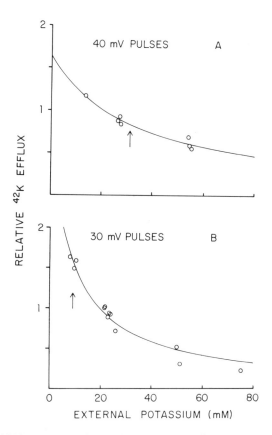

FIG. 3. Relative inhibition, by potassium in the seawater, of ^{42}K efflux through K channels of squid giant axon, for two different clamping pulse sizes. For each of the two pulse sizes—40 mV **(A)**, 30, mV **(B)**—the fluxes were scaled to unity at 20 mM external K. The solid line through the data is a least-squares fit to Eq.[2] (see text). The equilibrium constant for $[K]_o$ inhibition is shown by the arrow in this figure and is clearly voltage dependent.

more complex if necessary by including more binding sites and by allowing some sites to be occasionally empty.

Since each collision transfers a unit charge, the knock-on model predicts that membrane current will be described by the independence principle (11). Table 2 shows the results of tests of the independence principle on K currents. Although there is some variability, no consistent trend appears and the mean of all data are in agreement with the K current's behaving in an independent manner.

More experimental and theoretical work is necessary before the ionic permeation mechanism in K channels will be known. The data presented here are qualitatively consistent with a single-site knock-on model for this process. However, the strong voltage dependence of the inhibitory equilibrium constant suggests there is more than just one site.

TABLE 2. *Current ratios and independence*

Axon	$[K]_o$ (mM)	V_m (mV)	Independence prediction	Measured	Measured / prediction
6–15B	5	−48	1.3	1.87	1.44
6–15C	5	−58	1.18	1.97	1.67
6–16A	5	−38	1.28	1.26	.98
	50	−38	0.53	0.55	1.04
	5	−48	1.46	1.5	1.03
6–16B	50	−38	0.44	0.41	0.93
6–17A	50	−38	0.65	0.52	0.8
6–17D	5	−48	1.4	1.5	1.07
6–18A	75	−48	−0.34	−0.42	1.24
6–18B	50	−58	−2.08	−1.63	0.78
Mean ± SD					1.1 ± 0.28

REFERENCES

1. Armstrong, C. M. (1975): Potassium pores of nerve and muscle membranes. In: *Membranes: A Series of Advances*, Vol. 3, edited by G. Eisenman. Dekker, New York.
2. Adelman, W. J., Jr., Palti, Y., and Senft, J. P. (1973): Potassium ion accumulation in a periaxonal space and its effect on the measurement of potassium ion conductance: *J. Membr. Biol.*, 13:387–410.
3. Baker, P. F., Hodgkin, A. L., and Shaw, T. I. (1962): Replacement of the axoplasm of giant nerve fibers with artificial solutions. *J. Physiol.*, 164:330–337.
4. Begenisich, T., and Cahalan, M. (1978): Studies of ion permeation through Na channels. (*In preparation.*)
5. Begenisich, T., and De Weer, P. (1977): Ionic interactions in the potassium channels of squid giant axons. *Nature*, 269:710–711.
6. Bezanilla, F., Rojas, E., and Taylor, R. E. (1970): Time course of the sodium influx in squid giant axon during a single voltage clamp pulse. *J. Physiol.*, 207:151–164.
7. Ehrenstein, G., and Gilbert, D. L. (1966): Slow changes of potassium permeability on the squid giant axon. *Biophys. J.*, 6:533–566.
8. Hille, B. (1975): Ionic selectivity, saturation, and block in sodium channels: A four-barrier model. *J. Gen. Physiol.*, 66:535–560.
9. Hille, B. (1975): Ionic selectivity of Na and K channels of nerve membranes. In: *Membranes: A Series of Advances*, Vol. 3, edited by G. Eisenman. Dekker, New York.
10. Hladky, S. B., and Harris, J. D. (1967): An ion displacement membrane model. *Biophysics. J.*, 7:535–543.
11. Hodgkin, A. L., and Huxley, A. F. (1952): Currents carried by sodium and potassium ions through the membrane of the giant axon of *Loligo. J. Physiol.*, 116:449–472.
12. Hodgkin, A. L., and Keynes, R. D. (1955): The potassium permeability of a giant nerve fibre. *J. Physiol.*, 128:61–88.
13. Lauger, P. (1973): Ion transport through pores: A rate theory analysis. *Biochim. Biophys. Acta*, 311:423–441.
14. Yeh, J. Z., Oxford, G. S., Wu, C. H., and Narahashi, T. (1976): Dynamics of aminopyridine block of potassium channels in squid axon membrane. *J. Gen. Physiol.*, 68:517–536.

Membrane Transport Processes, Volume 3,
edited by C. F. Stevens and R. W. Tsien.
Raven Press, New York, © 1979.

Non Independence and Selectivity in Sodium Channels

Ted Begenisich and *Michael Cahalan

*Department of Physiology, University of Rochester School of Medicine and Dentistry,
Rochester, New York 14620; and *Department of Physiology, University of California
College of Medicine, Irvine, California 92717*

One of the main problems in studies of ion permeation is to determine how certain membrane pores discriminate between ions. This issue has been extensively discussed, especially by Hille (6). In this work, Hille describes various ways of measuring selectivity. Of these, one that has received considerable attention is to determine ionic permeability ratios for a particular ionic pathway from the zero-current or reversal potential:

$$V_{rev} = RT/F \; ln \frac{P_X \; [X]_o}{P_Y \; [Y]_i} \tag{1}$$

where V_{rev} is the potential at which the net current through the pathway vanishes, R is the gas constant, F is Faraday's constant, and T is the absolute temperature. RT/F at room temperature is about 25 mV. Brackets denote the concentration (actually the chemical activity) of ions in solution. This equation was written for the biionic case where only ion X is on one side (the outside) of the membrane ($[X]_i = 0$) and only ion Y is on the other ($[Y]_o = 0$). Then, if the ionic concentrations $[X]_o$, and $[Y]_i$ are known and V_{rev} is measured, the ionic permeability ratio P_X/P_Y can be computed. From this equation, one would expect that it would not matter what particular values $[X]_o$ and $[Y]_i$ had as long as they were known. Unfortunately, for certain membrane channels, this is not the case.

In 1965, Chandler and Meves (3) studied the selectivity properties of the Na channel of squid giant axons. Although P_{Na}/P_K was invariant under a wide variety of conditions including internal pH and internal anion changes, different values of the permeability ratio were obtained when the internal K concentration was changed. We (1,2) have extended these observations to internal Na, NH_4, Rb, and Cs and external Na and NH_4. We found that all the internal permeant ions tested except for Na affect permeability ratios. Neither external Na nor NH_4 altered the permeability ratios. So, the effect is asymmetric with respect to ion type and membrane surface.

Ebert and Goldman (4) found similar results using *Myxicola* giant axons,

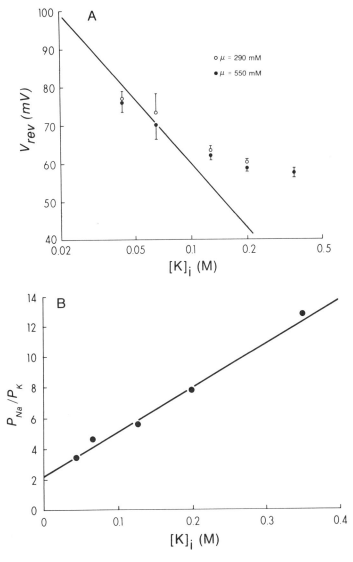

FIG. 1. A: Squid axon sodium channel reversal potential as a function of internal potassium activity. External solution was sodium-containing artificial (K-free) seawater. The internal solution contained various concentrations of impermeant tetramethylammonium ions to maintain the ionic strengths shown. **B:** P_{Na}/P_K calculated from text Eq. [1] for the 550 mM ionic strength data of **A.**

but their work was restricted to internal K, Rb, and Cs. Meyers and Haydon (9), Eisenman et al. (5), and Sandblom et al. (10) found that gramicidin channels in lipid bilayers behave as though selectivity were a function of the permanent ion concentration.

Some of our results (1) are shown in Fig. 1. Part A of Fig. 1 shows the squid axon Na channel reversal potential as a function of internal K activity for two internal ionic strengths (550 and 290 mM). The ionic strength was changed with the impermeant cation tetramethylammonium. There is no significant difference between the results at the two ionic strengths. The solid line in Fig. 1A is the expected result from Eq. [1] if P_{Na}/P_K were constant; the actual results differ significantly from this prediction.

Figure 1B shows P_{Na}/P_K determined from Eq. [1] plotted against internal K activity. From this, it appears that P_{Na}/P_K is approximately a linear function of $[K]_i$. Control experiments showed that any contribution of anions to the reversal potential is negligibly small (1). Also, the effect is asymmetric: changing Na on the outside of the fiber does not change P_{Na}/P_K. This is an important and useful finding since it puts severe constraints on possible models of ion permeation.

One could choose to model this effect by saying that ion permeability ratios are simply functions of ionic concentration for some ions on one side of the membrane. Physically, such a model could come from the binding of an ion to a particular location that, through some (unspecified) mechanism, changes the structure of the ion pathway so that the selectivity is altered. Such a model may indeed be correct. However, we would like to investigate other possible models. In particular, we would like to see if models for ionic permeation based on the principles of Eyring rate theory can produce these effects without explicitly making permeabilities functions of concentration.

MODEL I

The first such model to try is a one-ion–one-site pore. This model has been discussed by Läuger (8), Hille (7), and others, but is presented here as a convenient starting point.

We consider the case where there are only two ions, say Na and K, present. We also consider that there is a site within the membrane, and this site can be either empty, denoted by S, or filled with either a K or a Na ion, denoted by KS or NaS. For a Na ion to get to the site from the outside of the membrane, it must cross an energy barrier, G_{12}; to get from the site to the inside it must cross an energy barrier, G_{23}. The site is a minimum of energy and is called G_2. Figure 2 shows the Na and K energy diagrams for this model. The equations describing the reaction between these ions and the site are as follows:

$$Na_o + S \underset{\kappa_{21}}{\overset{\kappa_{12}}{\rightleftarrows}} NaS \underset{\kappa_{32}}{\overset{\kappa_{23}}{\rightleftarrows}} Na_i + S \qquad [2a]$$

$$K_o + S \underset{l_{21}}{\overset{l_{12}}{\rightleftarrows}} KS \underset{l_{32}}{\overset{l_{23}}{\rightleftarrows}} K_i + S \qquad [2b]$$

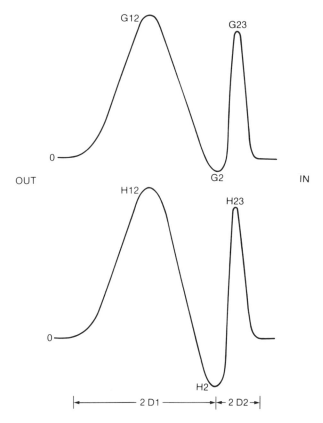

FIG. 2. Arbitrary two-barrier-one-site energy profiles for Na and K ions in a Na channel. K^+ barriers are denoted by H_{12} and H_{23}. D's represent electrical distance through the membrane. $2D1 + 2D2$ must equal 1. Rate constants for crossing barrier G_{12}, etc. (ν is the frequency factor that has been included in the nonvoltage-dependent part of the rate constants denoted by superscript zero):

For Na

$$\kappa_{12} = \nu\exp\left[-(G_{12} + VFD_1)/RT\right] = \kappa_{12}^0\exp\left(-VFD_1/RT\right)$$
$$\kappa_{21} = \nu\exp\left[-(G_{12} - G_2 - VFD_1)/RT\right] = \kappa_{21}^0\exp\left(VFD_1/RT\right)$$
$$\kappa_{23} = \nu\exp\left[-(G_{23} - G_2 + VFD_2)/RT\right] = \kappa_{23}^0\exp\left(-VFD_2/RT\right)$$
$$\kappa_{32} = \nu\exp\left[-(G_{23} - VFD_2)/RT\right] = \kappa_{32}^0\exp\left(VFD_2/RT\right).$$

For K

$$l_{12} = \nu\exp\left[-(H_{12} + VFD_1)/RT\right] = l_{12}^0\exp\left(-VFD_1/RT\right)$$
$$l_{21} = \nu\exp\left[-(H_{12} - H_2 - VFD_1)/RT\right] = l_{21}^0\exp\left(VFD_1/RT\right)$$
$$l_{23} = \nu\exp\left[-(H_{23} - H_2 + VFD_2)/RT\right] = l_{23}^0\exp\left(-VFD_2/RT\right)$$
$$l_{32} = \nu\exp\left[-(H_{23} - VFD_2)/RT\right] = l_{32}^0\exp\left(VFD_2/RT\right).$$

The differential equations for NaS and KS are:

$$\frac{d}{dt}[NaS] = [S]\,([Na]_o\kappa_{12} + [Na]_i\kappa_{32}) - [NaS]\,(\kappa_{21} + \kappa_{23})$$

$$\frac{d}{dt}[KS] = [S]\,([K]_o l_{12}[K]_i l_{32}) - [KS]\,(l_{21} + l_{23})$$

The time dependence of redistribution of ions within the pore from any imposed perturbation in the system is assumed to be fast compared to our measuring apparatus. We, therefore, solve the steady-state case where $d/dt\,[KS] = d\,[NaS]/dt = 0$. Also, we solve here the biionic case of $[Na]_i = [K]_o = 0$ and add the conservation equation on the total number of sites, S_T: $[S] + [KS] + [NaS] = S_T$.

The Na influx/site, f_{Na}, is given by $[NaS]\kappa_{23}$, and the K efflux/site, f_κ, is given by $[KS]l_{21}$. The final solutions are:

$$f_{Na} = [Na]_o\,\kappa_{23}\kappa_{12}\,(l_{21} + l_{23})/\text{DENOM} \qquad [3a]$$

$$f_K = [K]_i\,l_{32}l_{21}\,(\kappa_{21} + \kappa_{23})/\text{DENOM} \qquad [3b]$$

$$\text{DENOM} = [Na]_o\kappa_{12}(l_{21} + l_{23}) + [K]_i\,l_{32}\,(\kappa_{21} + \kappa_{23})$$
$$+ (\kappa_{21} + \kappa_{23})(l_{21} + l_{23})$$

The rate constants are determined using Eyring rate theory and are shown in Fig. 2 legend. Substituting these expressions for the rate constants into Eqs. [3a] and [3b] and solving for the potential at which the fluxes are equal, we find:

$$V_{rev} = RT/F\,ln\frac{P_{Na}[Na]_o}{P_K[K]_i} \qquad [4]$$

where

$$P_{Na} = \frac{\kappa_{23}^o\,\kappa_{12}^o}{\kappa_{21} + \kappa_{23}} \quad P_K = \frac{l_{32}^o\,l_{21}^o}{(l_{21} + l_{23})}$$

Equation [4] is just Eq. [1], except now the permeabilities are explicitly functions of potential. The voltage dependence can be adjusted to produce the results of Fig. 1, but can not simultaneously produce the asymmetry with regard to external Na as described above. Extending the model to include more sites and more barriers will not change this conclusion—as long as no more than one ion is allowed to occupy the pore at any time.

MODEL II

The second model to be discussed allows more than one ion at a time in the pore. For the general case of even only two sites, the mathematics are extremely tedious. However, a simple way to include more than one ion at a time is to allow one ion to knock another ion off the site. Then in addition to the reactions in Eqs. [2a] and [2b], we have:

$$\mathrm{Na}_o + \mathrm{K}S \underset{m_2}{\overset{m_1}{\rightleftarrows}} \mathrm{Na}S + \mathrm{K}_i \qquad [5a]$$

$$\mathrm{Na}_o + \mathrm{Na}S \underset{n_2}{\overset{n_1}{\rightleftarrows}} \mathrm{Na}S + \mathrm{Na}_i \qquad [5b]$$

$$\mathrm{K}_o + \mathrm{K}S \underset{q_2}{\overset{q_1}{\rightleftarrows}} \mathrm{K}S + \mathrm{K}_i \qquad [5c]$$

As in the first model, the steady-state solutions are obtained. The expressions for Na and K fluxes are found to be:

$$f_{\mathrm{Na}} = [\mathrm{Na}]_o([\mathrm{Na}]_o n_1 + \kappa_{23})\{[\mathrm{K}]_i m_1 l_{32} + \kappa_{12}([\mathrm{Na}]_o m_1 + l_{21} + l_{23})\}/\mathrm{DENOM} \qquad [6]$$

$$f_{\mathrm{K}} = [\mathrm{K}]_i([\mathrm{K}]_i q_2 + l_{21})\{[\mathrm{Na}]_o m_2 \kappa_{12}\ l_{32}([\mathrm{K}]_i m_2 + \kappa_{21} + \kappa_{23}\}/\mathrm{DENOM}$$

$$\mathrm{DENOM} = [\mathrm{K}]_i^2 l_{32} m_2 + [A]_o^2 \kappa_{12} m_1 + [\mathrm{K}]_i\{l_{32}(\kappa_{21} + \kappa_{23}) + m_2(l_{21} + l_{23})\}$$
$$+ [\mathrm{Na}]_o\{\kappa_{12}(l_{21} + l_{23}) + m_1(\kappa_{21} + \kappa_{23})\}$$
$$+ [\mathrm{Na}]_o[\mathrm{K}]_i(\kappa_{12} m_2 + m_1 l_{32}) + (l_{21} + l_{23})(\kappa_{21} + \kappa_{23})$$

Like the experimental observations, this model shows apparent concentration-dependent permeabilities. In fact, if $m_2 = n_1 = q_2 = 0$ then $P_{\mathrm{Na}}/P_{\mathrm{K}} = d[\mathrm{K}]_i + b$ where a is a function of voltage but not concentrations and b is a function of voltage and $[\mathrm{Na}]_o$. This is just the type of relationship shown in Fig. 1B.

To test this model quantitatively, the data of Fig. 1A have been replotted in Fig. 3A along with the predictions of Eqs. [6] (*solid line*) with $m_2 = n_1 = q_2 = 0$. This equation certainly fits the data better than Eq. [1]. If the restrictions on the rate constants are removed, a similar but slightly poorer fit is obtained. The asymmetric effect of external Na can be accounted for by this model. The asymmetry arises from the asymmetric rate constants, barrier heights, and well depths. So this simple version of a two-ion pore can account for the observations of ionic concentration-dependent permeability ratios.

The predictions of this model differ from those of the one-ion pore model in another respect: there is no saturation of the ionic current as the concentration of the permeant ions is raised. Figure 3B shows the calculated current-versus-concentration curve for the two ion-one site pore (model II) for the case where there is only one permeant ion on the inside of the cell. This relationship obviously does not saturate. Also shown is the same relationship for the one-ion pore (model I) which does show saturation. Since the squid axon Na channel shows saturation (2), the present form of the two-ion pore has limited utility for this ionic pathway. However, to account for concentration-dependent permeabilities, some form of multiple ion pores must be considered.

Another form of multiple-ion pores is given by Eisenman et al. (5). This model also shows apparent concentration-dependent selectivities.

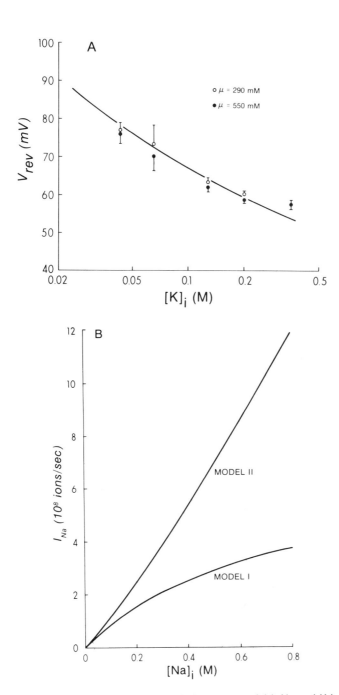

FIG. 3. A: Data of Fig. 1A replotted. Solid line is from text model I. Na and K barriers were 8, 7 RT and 8.7, 7.7 RT. D1, D2 were 0.3 and 0.2. R is the gas constant; T is the absolute temperature. RT at 10°C is approximately 0.56 Kcal/mole. **B:** Na current versus internal Na concentration ($[Na]_o \doteq 0$). Solid lines are from text models I and II. Na parameters for each model are the same as above except that model I has no interaction.

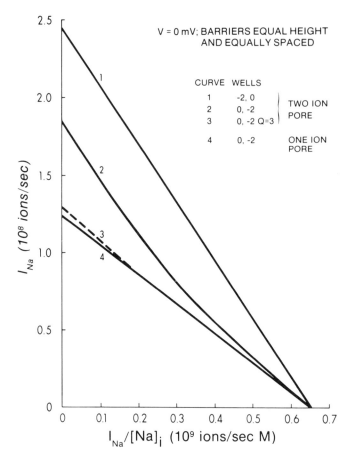

FIG. 4. Eadie–Hofstee plot of sodium current vs internal Na concentration ($[Na]_o = 0$). Barrier heights were all 8 RT, and well depths are given in RT units.

Since it is clear that multiple-ion pore models will play an increasingly larger role in simulating ion permeation processes, it is useful to examine some of the properties of such models. For the purpose of simplifying the necessary calculations, we consider a three-barrier, two-binding-site pore. We allow either or both sites to be filled by ions of only one type (say Na). Furthermore, we treat the case where this ion is only on one side of the membrane (e.g., the inside).

The equations describing this model are similar to, but more complicated than, those for models I and II. The derivation of these equations can be done in the same manner as that above or they can be derived by the use of King-Altman diagrams as described by Hille (7).

Some computations for the three-barrier model are plotted in Fig. 4 as the current through the pore as a function of the current divided by the ionic concentration. This is called an Eadie–Hofstee plot and is often used to describe enzyme kinetics and has also been used to describe ion permeation processes (5). A straight line on an Eadie–Hofstee plot usually represents a single binding site and the slope of the line is $-K_m$, the equilibrium constant. For all the curves in Fig. 4, the energy barriers were of equal height and were equally spaced. In curve 1, there is a deep well (binding site) toward the outside of the membrane—away from the side containing the permeant ions. Curve 1 is a straight line.

Curve 2 is a similar plot but now with the deep well close to the side of the membrane where the permeant ions are; there are now two slopes. Curve 4 is the same as 2 except that it was computed with the restriction that no more than one ion could occupy this two-site pore; there is only one slope. So even with two binding sites, only in certain situations do two slopes appear in the Eadie–Hofstee plot.

The two-ion pore model can also include interactions between the ions. It might reasonably be expected that two cations would mutually repel each other when contained within a channel only 25 to 40 Å long. If the dielectric constant within the pore were known, the degree of repulsion as a function of separation could be calculated. This value is not known, but an assumed dielectric constant of 20 leads to a repulsive potential of 72 mV for the ions spaced by 10 Å. This is equivalent to approximately three times thermal energy. Even though this simple calculation may not be correct for many reasons, including the fact that there may be discrete charges within the pore, it serves as a reasonable magnitude for computational purposes.

Curve 3 of Fig. 4 was computed using the same conditions as for curve 2, except that ionic interaction is now included. The interaction was included by decreasing by 3 RT the energy barrier for an ion to leave the pore when both sites are filled. Also, when one ion is in the pore, the barrier for another ion to enter is raised by 3 RT. It can be seen from the figure that this interaction causes the two-ion pore results to approximate those of the one-ion pore. This is to be expected since the presence of mutual repulsion reduces the probability of having more than one ion in the pore at any time.

The experimental data on ion permeation seem rather complicated and include saturating current-concentration relations and concentration-dependent permeabilities. The mathematical models for these processes may also be complicated or at least tedious. Our computations show that to account for the experimental observations such models must provide for at least two ions at a time in the pore and probably must also include interactions between the ions. The calculations of Fig. 4 show that the presence of only one slope in an Eadie–Hoffstee plot does not rule out the presence of a second binding site and that a two-ion model with ionic interactions cannot always be distinguished from a one-ion pore. More experiments and calculations will need to be done, perhaps including radioactive tracer flux measurements, before a clear picture of even the most basic properties of ionic pores can be determined.

REFERENCES

1. Cahalan, M., and Begenisich, T. (1976): Sodium channel selectivity: dependence on internal permeant ion concentration. *J. Gen. Physiol.,* 68:111–125.
2. Cahalan, M., and Begenisich, T. (1978): Studies of ion permeation through Na channels *(in preparation).*
3. Chandler, W. K., and Meves, H. (1965): Voltage clamp experiments on internally perfused giant axons. *J. Physiol.,* 180:788–819.
4. Ebert, G. A., and Goldman, L. (1976): The permeability of the sodium channel in *myxicola* to the alkali cations. *J. Gen. Physiol. [Lond.],* 68:327–340.
5. Eisenman, G., Sandblom, J. P., and Neher, E. (1977): Ionic selectivity, saturation, binding and block in the gramicidin A channel: a preliminary report. In: *Metal-Ligand Interactions in Organic Chemistry and Biochemistry* (Part 2), edited by B. Pullman and N. Goldblum, pp. 1–36. D. Reidel Publishing Company, Dordecht-Holland.
6. Hille, B. (1975): Ionic selectivity of Na and K channels of nerve membranes. In: *Membranes: A Series of Advances. Vol. 3,* edited by G. Eiseman, Marcel Dekker, N.Y.
7. Hille, B. (1975): Ionic selectivity, saturation and block in sodium channels: A four-barrier model. *J. Gen. Physiol.,* 66:535–560.
8. Läuger, P. (1973): Ion transport through pores: A rate theory analysis. *Biochim. Biophys. Acta,* 311:423–441.
9. Meyers, V. B., and Haydon, D. A. (1972): Ion transfer across lipid membranes in the presence of gramicidin A. II. The ion selectivity. *Biochim. Biophys. Acta,* 274:313–322.
10. Sandblom, J., Eisenman, G., and Neher, E. (1977): Ionic selectivity, saturation and block in gramicidin A channels: I. Theory for the electrical properties of ion selective channels having two pairs of binding sites and multiple conductance states. *J. Membr. Biol.,* 31:383–417.

Membrane Transport Processes, Volume 3,
edited by C. F. Stevens and R. W. Tsien.
Raven Press, New York, © 1979.

Modulation of Conductance at the Neuromuscular Junction

Vincent E. Dionne

Department of Physiology and Biophysics, University of Vermont, College of Medicine, Burlington, Vermont 05401

Nervous electrical activity is commonly coupled to other cells by a chemical transduction process that occurs at specialized intracellular junctions termed synapses. In this chapter I discuss one specific question about the mechanism of synaptic transmission, that is, How many different types of "postsynaptic ionic channels" are activated on stimulation? The following paragraphs serve to introduce and motivate the question.

The processes of chemical transduction have been studied by electrophysiologists most thoroughly at the twitch neuromuscular junction of frogs where the chemical transmitter is acetylcholine. Neuromuscular synapses in the frog are relatively large and easily accessible, favoring the animal's use. The normal sequence of neuromuscular synaptic events begins with a nerve action potential propagating into the presynaptic nerve ending. There the action potential causes the nerve to depolarize, and within milliseconds, acetylcholine is released from the ending onto regions of the postsynaptic muscle membrane tightly apposed to the nerve. In these regions of the muscle, special receptors exist that bind the chemical transmitter and trigger a small, brief increase in the membrane conductance. Normally the nerve releases sufficient acetylcholine to produce a conductance change large enough that the synaptic region of the muscle cell depolarizes beyond threshold to initiate a muscle action potential. This electrical activity is then propagated along the muscle surface and into the transverse tubular network to produce contraction. The general sequence of synaptic events at the neuromuscular junction is believed representative of that at most chemical synapses. However, specific molecular characteristics of synaptic transmission are expected to differ, especially where the chemical transmitter is not acetylcholine.

The conductance change mediated by acetylcholine in the postsynaptic membrane is thought to involve transmembrane channels that can open and close. The idea that channels constitute the ion translocation structure is based on the magnitude of current carried by each ion translocator. At -90 mV a net current of approximately 1.4×10^7 ions/sec is carried by the individual channels, the sum of the one-way currents being even greater. Since diffusion could supply ions to the channel access with a rate of about 10^8 ions/sec, it seems unlikely that the translocation device could be anything more complicated than a pore and still be able to function with the observed rate.

When acetylcholine induces postsynaptic channels to open, sodium and potassium ions flow down their electrochemical potential gradients to locally depolarize the muscle membrane. These cations carry most of the current through the synaptic channels. In the nerve and elsewhere in the muscle, these same ions are responsible for the action potential. The molecular basis of the action potential is two distinctly separate populations of channel types—the Na channel and the K channel. These are voltage-activated channels that yield pharmacologically separable currents with different temporal and voltage-dependent behaviors. It becomes reasonable then to consider whether the acetylcholine-activated end-plate current consists of sodium and potassium ions moving through a population of channels in which they are both permeable or whether these ions cross the membrane through distinguishable ion-selective channels such as those for the nerve and muscle action potentials, but chemically gated.

In the remainder of this chapter I discuss this question. For brevity, I commonly refer to the alternative channel mechanisms as the single-channel model and the two-channel model. These titles are not very accurate, but they serve to distinguish the separate ideas. After the historical section that follows, I introduce a sensitive test that potentially allows us to identify which model actually describes the molecular mechanism responsible for the end-plate conductance change. There I detail the limitations placed by the test on our description of the channel mechanism.

HISTORICAL BACKGROUND

Takeuchi and Takeuchi (14) published the first studies of voltage clamped end-plate currents. By altering extracellular ion concentrations, they showed that sodium and potassium ions were responsible for the current that caused the end-plate potential. To model their results, they employed conceptually separate sodium and potassium conductances that were altered by acetylcholine. However, they presented no evidence whether the sodium and potassium conductance mechanisms were physically independent.

The proposal that there are separate sodium and potassium channels at the end plate was made by Maeno (7). He noted that the local anesthetic procaine caused nerve-evoked end-plate currents to decay with a two-component time course at negative voltages (Fig. 1). However, at the estimated equilibrium potential for sodium ions, V_{Na} near $+40$ mV, where the entire current should be carried by potassium, the end-plate currents decayed as a single component, stimulating the suggestion that there are separate conductances for sodium and potassium and that the sodium conductance mechanism specifically is susceptible to alteration by procaine.

Gage and Armstrong (4) measured the time course of miniature end-plate currents at voltages near the equilibrium values for sodium and potassium. A

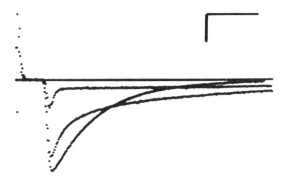

FIG. 1. Shown here are the qualitative effects produced by increased concentrations of local anesthetics on the decay time course of end-plate currents. These traces illustrate the change in decay produced by the drug QX-222 at two concentrations relative to the decay in normal Ringers. The trace with the most negative peak was recorded in normal Ringers. With 0.1 mM QX-222, the peak current was slightly smaller and the time course altered. QX-222 at 0.5 mM reduced the peak to approximately 30% of normal. (V = −125 mV; calibration bars: 200 nA, 5 msec). In general, the drugs caused a rapid initial decline of current followed by a slowed decay phase. Similar effects are produced by procaine and other local anesthetics. Ruff (10) studied the kinetics of QX-222 action using both miniature end-plate currents and current fluctuations; the kinetics suggested a mechanism of action such as a transient blockage of the open channels by the drug. (From ref. 1, with permission.)

miniature end-plate current is produced by the spontaneous release of one quantum of acetylcholine from the nerve; the quantum of acetylcholine contains about 10^4 molecules and causes several hundred end-plate channels to open simultaneously. Presumably at V_K only sodium ions carry the end-plate current, whereas at V_{Na} only potassium ions carry net current in the channels. They observed that at the equilibrium voltages the time courses of the miniature end-plate currents were different. The miniature end-plate currents near the sodium equilibrium potential decayed faster than those at the potassium potential. This result was interpreted as evidence for separate sodium and potassium channels opened by acetylcholine, the conductance changes having similar but clearly different time courses.

At about the same time, Steinbach (11) presented an analysis of xylocaine-altered end-plate potentials. This local anesthetic, presumably acting in a manner similar to that of procaine, produced a two-component decay that was consistent with a drug-induced alteration of normally activated channels of the type through which all the end-plate current ions could pass. Altered channels required only a reduced conductance to produce the observed local anesthetic effects.

Kordas (5) also argued against the two-channel hypothesis, pointing out that if separate channels for sodium and potassium were opened with slightly different time courses, as suggested, then no true reversal potential could exist. Instead, near the apparent reversal potential, one should observe first a fast outward, predominantly K current followed by an inward Na current as the slower Na channels remained open and the potassium conductance decayed to zero. The end-plate current recorded at the reversal potential would be biphasic. This behavior should be made more obvious following treatment with procaine, which would presumably slow the Na channel response while leaving the K current unchanged (6). Neither with nor without procaine could Kordas observe biphasic end-plate currents near the reversal potential. His end-plate currents always reversed without complication. However, Maeno et al. (8) reported observing occasional biphasic end-plate currents near the reversal potential in glycerol-treated end-plates. These currents were small and were often obscured by noise.

Kordas' view was supported by Magleby and Stevens (9) who studied the voltage dependence of the time course of end-plate current decay. They pointed out that the single exponential behavior of the end-plate current, which they observed at all voltages, was inconsistent with proposals of two separate, voltage-independent conductance changes. They were careful to make clear that their data did not address directly the question of whether sodium and potassium ions pass through the same or separate end-plate channels. Indeed, none of the experiments that I have surveyed here bears directly on this question. Rather these reports all describe observations that can be interpreted in several ways. None conclusively distinguishes between the basic aspects of either a population of common channels for sodium and potassium movement or separate populations of Na channels and K channels at the end plate.

A TEST OF THE END-PLATE CHANNEL MECHANISM

It is possible to formulate a test of whether the individual end-plate current ionic components move through commonly gated channels or kinetically independent, ion-selective channels. This test is described below. Note here that the measurements allow us to distinguish whether an acetylcholine-activated gating mechanism opens to allow all the ions in the end-plate current through the same membrane channel structure or acetylcholine activates independent, separately gated ion-selective channels. These are the alternatives that I have called the, respectively, single-channel and the two-channel models. The test addresses directly the nature of the gating of end-plate channels. It does not tell us the nature of ion-selectivity within the channel. For example, we see below that the experimental evidence favors a single-channel mechanism. However, the experiments do not allow us to decide, for instance, whether the same interaction sites within a channel are sensed by both sodium and potassium ions while in transit. A treatment of channel conductance is contained in the chapter by Lewis and Stevens (6a) in this volume.

The test of whether there are kinetically independent channels for sodium and potassium or commonly gated channels at the end plate rests on a simple quantitative description of the end-plate current. There is one inherent assumption—that the channels open and close independent of one another. The assumption of independence is necessary to compute expressions for end-plate current variance. If there were channels permeable to both sodium and potassium whose gates were not independent, then the descriptions here would be inappropriate.

That is, we can imagine several possible relations between the channels and their gates: (a) channels permeable to both Na and K, each with one gating mechanism, independent of all other channels; (b) channels as in (a) but with coupled gates; (c) separate Na and K channels with separate, independent gates; and (d) either partially or fully separate Na and K channels with coupled gates. The independence assumption means that the quantitative descriptions to be developed here apply to (a) and (c) above, but not to (b) and (d). However, since the test does not distinguish factors having to do with ionic selectivity, relations (a) and (d) will be indistinguishable. Thus, the assumption of independence among channels does not severely limit the applicability of the test although it is critically necessary for the quantitative formulation.

Let us first suppose that the ions of the end-plate current move through a population of identical, independent channels whose gates are triggered by acetylcholine molecules. If we were to apply acetylcholine to the channels, some of them would open. If the voltage across the membrane were maintained constant (as with a voltage clamp), we would observe a mean end-plate current given by

$$\bar{\mu}_1 = \gamma \, pN(V - V_r) \qquad [1]$$

where the subscript 1 indicates a population of end-plate channels of the single-channel type, γ is the mean single-channel conductance, N is the number of channels available at the end plate, p is the probability that these channels open, V_r is the end-plate current reversal potential (the voltage at which the current changes sign), and V is the membrane potential. This equation simply expresses Ohm's law: γpN is the total conductance change induced by the acetylcholine, the product of the number of open channels, and the mean single open channel conductance, and $(V - V_r)$ is the driving potential.

The actual value of μ_1 fluctuates around the mean value $\bar{\mu}_1$ if the application of acetylcholine is prolonged. These fluctuations arise because the number of open channels varies from instant to instant although hovering about an average value pN. This is a reflection of the short channel lifetime, only a few milliseconds, which necessitiates the continual opening of channels to maintain the mean end-plate current. Careful analysis of the current fluctuations can reveal aspects of channel behavior not normally measurable (12,13). Possibly the simplest way to characterize a fluctuating variable is with its variance σ^2, which contains information about the magnitude of the fluctuations around their mean value. The variance of the end-plate current is the average value that the square of the quantity $(\mu_1 - \bar{\mu}_1)$ assumes. With the assumption of independence among channels, it can be written in terms of the probability of opening a single channel, p, as

$$\sigma_1^2 = \gamma^2 \, p(1 - p)N(V - V_r)^2 \qquad [2]$$

Equations [1] and [2] for the mean end-plate current and its variance provide a basic quantitative description for a population of independent, identical channels in which all the ions are simultaneously gated. As an alternative, we can write similar expressions with the assumption that there are separate sodium and potassium channels at the end plate. Although it appears that other cations such as calcium can pass through the end-plate channels, we ignore here their small contribution to the current. The concept of separate channels for each ion could be extended to include a Ca channel and channels for other ions, but it will be apparent that the conclusions drawn using the simpler two-channel model will be unchanged by the addition of more ion-selective channels.

Let us now suppose that the ions of the end-plate current pass through separate populations of sodium- and potassium-selective, independent channels that are activated by acetylcholine. The application of acetylcholine causes some channels of both types to open, and a description of the mean end-plate current at constant voltage is given by

$$\bar{\mu}_2 = \gamma_{Na} p_{Na} N_{Na} (V - V_{Na}) + \gamma_K p_K N_K (V - V_K) \qquad [3]$$

This expression is similar to Eq. [1], but contains separate terms for the sodium and potassium currents. The symbols have meanings similar to those described; V_{Na} and V_K are the Nernst potentials for sodium and potassium. The expression for the variance also has two terms, one each for sodium and potassium, because the channels are assumed independent;

$$\sigma_2^2 = \gamma_{Na}{}^2 p_{Na} (1 - p_{Na}) N_{Na} (V - V_{Na})^2 + \gamma_K{}^2 p_K (1 - p_K) N_K (V - V_K)^2. \quad [4]$$

The most significant difference between the equations expressing the current mean and variance for the two models is the value of the equilibrium voltages that enter. Normal end-plate currents exhibit a reversal potential, V_r, that is within a few millivolts of zero. This value is clearly different from the Nernst potentials for sodium ($V_{Na} \simeq +40$ mV) and potassium ($V_K \simeq -100$ mV) in frog muscle. This difference allows us to test between the alternative models.

To identify the correct model, one has only to determine the reversal potential and at that voltage measure the variance of the end-plate current. An inspection of Eq. [1] through [4] indicates what to expect. The single-channel model predicts that at $V = V_r$ both the mean current and its variance are zero, because the net driving potential is zero (Eq. [1] and [2]). In contrast, the separately gated two-channel model predicts a net zero current at $V = V_r$ because the individual Na and K currents are equal but opposite, and it predicts a nonzero variance. At the reversal potential the variance for the two-channel model is still described by two terms that are both positive and cannot cancel. Thus, the experimental test requires an evaluation of end-plate current variance at the reversal potential; the single-channel model predicts zero variance, whereas the two-channel model predicts a nonzero variance.

Measurements of the type described here have been reported by Dionne and Ruff (3) on frog twitch fiber end plates that were either untreated or exposed to procaine. In addition Ruff (10) reported measurements on the same preparation treated with the drug QX-222, a trimethyl derivative of lidocaine with local anesthetic-like properties at the end-plate, and Dionne and Parsons (2) have made measurements on untreated garter snake twitch fiber neuromuscular junctions. All these reports arrive at the same conclusion: at the reversal potential the end-plate current variance is zero, indicating that the ions that carry the end-plate current move through a singly gated channel structure, not through independent, separately gated ion-selective channels. An example of data from the snake appears in Fig. 2; it shows how the mean end-plate current and its variance change with voltage between −100 and +10 mV. At the apparent reversal potential of −1.5 mV the acetylcholine-induced variance is indistinguishable from zero. The reader should note that what is plotted here is the inducible variance, $\sigma^2_{acetylcholine}$, not the total current variance. The total current has two components, a holding current that maintains the muscle fiber membrane potential while the end-plate channels are closed and an acetylcholine-inducible end-plate current. It is assumed that the holding current in the absence of acetylcholine arises through conductance mechanisms independent of the end-plate channels. Then the holding current variance adds to the acetylcholine-

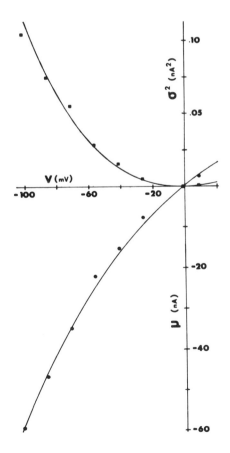

FIG. 2. Simultaneous measurements of mean end-plate current $\bar{\mu}$ and its variance σ^2 as a function of voltage. These data were recorded in a visually located end-plate on a twitch fiber in the garter snake (sp. *Thamnophis)* costocutaneous muscle. The bath did not contain TTX or any local anesthetic; temperature 16°C. The smooth curves were generated with equations derived from a quantitative treatment of the single-channel model described by Dionne and Ruff (3). Although this theory is not discussed in the text, the curves are included here to demonstrate the consistancy of the conclusion that a single homogeneous population of independent channels causes the end-plate conductance change.

induced variance to give the total variance, $\sigma^2_{total} = \sigma^2_{holding} + \sigma^2_{acetylcholine}$.

Two possibilities remain to be discussed that might qualify or negate the conclusion that the end-plate channels are similar, commonly gated structures through which all the ions move. First, if the probabilities of channel opening for both the Na channel and K channel were functions of voltage so that at $V = V_r$ both were identically zero, we would predict a zero variance at the reversal potential with both models. In this case the test I have outlined would not distinguish between the models. Second, how much variance should be expected of the two-channel model at the reversal potential? If the expected variance is so small as to be experimentally indistinguishable from the background, then it would not be observed although it existed in fact.

Consider the possibility that p_{Na} and p_K could be functions of voltage, both becoming zero at V_r. If this occurred, both the single-channel and two-channel models would predict zero variance at the reversal potential. Below I argue these probabilities cannot be zero at V_r and still describe the observed current-voltage relations. That is, the requirements that p_{Na}, $p_K = 0$ at $V = V_r$ while Eq. [3] describes the observed data are inconsistent. The inconsistency arises because the probability functions for sodium and potassium channels to open can vary between zero and one but cannot be negative. This property causes the model to predict a zero slope for the current-voltage curve at V_r while a nonzero slope is observed.

The voltage-dependent functional forms that p_{Na} and p_K may assume are limited, because at voltages on either side of V_r both channels must have positive probabilities that they will open to allow current flow. Thus these functions have a general character somewhat like a parabola that is centered at V_r and zero there, if they are not to be discontinuous. The important consequence of this voltage dependence is that the slopes of both p_{Na} and p_K will be zero at V_r. That is

$$\left. \frac{dp_{Na/K}}{dV} \right|_{V_r} = 0 \text{ and } p_{Na/K} \Big|_{V_r} = 0 \qquad [5]$$

Let us differentiate Eq. [3] with respect to voltage and evaluate the differential at V_r; this gives the slope of the current-voltage relation at the reversal potential. If the two-channel model is consistent with the observed behavior, this derivative should give reasonable values for the slope.
From Eq. [3]

$$\left. \frac{d\bar{\mu}_2}{dV} \right|_{V_r} = \gamma_{Na} N_{Na} \left[p_{Na} + (V - V_{Na}) \frac{dp_{Na}}{dV} \right]_{V_r}$$

$$+ \gamma_K N_K \left[p_K + (V - V_K) \frac{dp_K}{dV} \right]_{V_r}. \qquad [6]$$

Expression [6] reduces to zero in all cases, for, as stated in Eq. [5], the probability functions p_{Na} and p_K and their derivatives are zero at V_r. The two-channel model predicts a zero slope at the reversal potential. However, this is not observed; the current-voltage curves have nonzero slopes at V_r, as exemplified in Fig. 2. There the slope is approximately 320 nA/V at V_r, this data from a snake twitch muscle fiber. The same is true in frog (3). Hence the necessary voltage dependent functions for p_{Na} and p_K are inconsistent with the actually observed end-plate current. The probabilities of opening separate Na and K channels cannot be voltage-dependent functions that are zero at the observed reversal potential.

The second possible qualification to the common-gate conclusion is that the expected two-channel current variance at $V = V_r$ might be so low as to be indistinguishable from zero. We can estimate this expected variance in the same manner as Ruff (10). Suppose the separate channel conductances γ_{Na} and γ_K and the opening probabilities p_{Na} and p_K vary little with voltage near the reversal potential, say between -40 mV and $V_r \simeq 0$ mV. For $V_K \simeq -100$ mV and $V_{Na} \simeq +40$ mV Eq. [3] may be evaluated at V_r to give $\gamma_{Na}p_{Na}N_{Na}/\gamma_K p_K N_K = 2.5$. For small acetylcholine concentrations where p_{Na}, $p_K \ll 1$, the variance as a function of V can be written as

$$\sigma_2^2 = \gamma_K p_K N_K \left[2.5 \, \gamma_{Na} \, (V - V_{Na})^2 + \gamma_K \, (V - V_K)^2 \right]$$

With this we can compute the ratio of the expected variance at V_r to the observed variance at -40 mV as a function of the ratio of the single-channel conductances γ_{Na}/γ_K. For the range $0.1 \leq \gamma_{Na}/\gamma_K \leq 10$ we find

$$30\% < \frac{\sigma_2^2 \, (0 \text{ mV})}{\sigma_2^2 \, (-40 \text{ mV})} \leq 200\% \qquad [7]$$

But the observed ratio of the variances in this range is less than 2% for the data shown in Fig. 2, and depending on experimental technique, resolution typically ranges from a high of about 3% to well under 1%. Under particularly favorable conditions, Ruff (10) obtained values of 10^{-6} for this ratio. Finally, this range of the single-channel conductance ration (Eq. [7]) is entirely reasonable. If we analyzed the mean current-voltage in Fig. 2 on the assumption of separate channels for sodium and potassium, it would give $\gamma_{Na}/\gamma_K \sim 3$. Thus, the expected variance at V_r from the two-channel model should be large and easily observed; however, no variance is seen.

It must be concluded that the single-channel model provides an accurate description of the currents that flow through activated end-plate channels, whereas the two-channel model does not. It is worth recalling what specific concepts were embodied in the term "single-channel" model. The experimental techniques actually address the activation of end-plate channels and not the details of conduction within those channels. Hence, it is more accurate to say that the acetylcholine-activated conductance mechanism at the neuromuscular junction is a receptor-channel complex; each complex is individually and independently gated to simultaneously allow all ionic species in the end-plate current access to the channel. There are not independently activated channels for the individual ionic species at the end-plate.

ACKNOWLEDGMENT

This work was supported by grants NS 13581 and NS 12306 from the United States Public Health Service.

REFERENCES

1. Beam, K. G. (1976): A voltage-clamp study of the effect of two lidocaine derivatives on the time course of endplate currents. *J. Physiol. (Lond.)*, 258:279–300.
2. Dionne, V. E., and Parsons, R. L. (1977): Characteristics of the postsynaptic channel at snake neuromuscular junctions. *Biophys. J.*, 17:124a.
3. Dionne, V. E., and Ruff, R. L. (1977): End-plate current fluctuations reveal only one channel type at the frog neuromuscular junction. *Nature*, 266:263–265.
4. Gage, P. W., and Armstrong, C. M. (1968): Miniature end-plate currents in voltage-clamped muscle fibres. *Nature*, 218:363–365.
5. Kordas, M. (1969): The effect of membrane polarization on the time course of the end-plate current in frog sartorius muscle. *J. Physiol. (Lond.)*, 204:493–502.
6. Kordas, M. (1970): The effect of procaine on neuromuscular transmission. *J. Physiol. (Lond.)*, 209:689–699.
6a. Lewis, C., and Stevens, C. F. (1979): Mechanisms of ion permeation through channels in a postsynaptic membrane. *This volume.*
7. Maeno, T. (1966): Analysis of sodium and potassium conductances in the procaine end-plate potential. *J. Physiol. (Lond.)*, 183:592–606.
8. Maeno, T., Edwards, C., and Hashimura, S. (1971): Difference in effects on end-plate potentials between procaine and lidocaine as revealed by voltage-clamp experiments. *J. Neurophysiol.*, 34:32–46.
9. Magleby, K. L., and Stevens, C. F. (1972): A quantitative description of end-plate currents. *J. Physiol. (Lond.)*, 223:173–197.
10. Ruff, R. L. (1977): A quantitative analysis of local anaesthetic alteration of miniature end-plate currents and end-plate current fluctuations. *J. Physiol. (Lond.)*, 264:89–124.
11. Steinbach, A. B. (1968): A kinetic model for the action of xylocaine on receptors for acetylcholine. *J. Gen. Physiol.*, 52:162–180.
12. Stevens, C. F. (1972): Inferences about membrane properties from electrical noise measurements. *Biophys. J.*, 12:1028–1047.
13. Stevens, C. F. (1975): Principles and applications of fluctuation analysis: a nonmathematical introduction. *Fed. Proc.*, 34:1364–1369.
14. Takeuchi, A., and Takeuchi, N. (1960): On the permeability of end-plate membrane during the action of transmitter. *J. Physiol. (Lond.)*, 154:52–67.

Membrane Transport Processes, Volume 3,
edited by C. F. Stevens and R. W. Tsien.
Raven Press, New York, © 1979.

Mechanism of Ion Permeation Through Channels in a Postsynaptic Membrane

Carol A. Lewis and C. F. Stevens

*Department of Physiology, Yale University School of Medicine,
New Haven, Connecticut 06510*

The ideal situation in which to study the physical basis for ion movements through channels in biological membranes would meet a number of criteria. First, one would like to be able to control the ionic composition of the bathing medium, ideally on both sides of the membrane bearing the channel. Second, it should be possible to measure not only the reversal potential, i.e., the voltage at which no net current flows through the channel, but also currents that flow through individual channels when a driving voltage is present. Third, the channel should permit a variety of different ionic species to pass through, while at the same time showing different permeabilities for the various ion types. This last requirement facilitates the study of selectivity mechanisms; one would like to probe the channel "selectivity filter" with as large a variety of ions with different properties as possible. Finally, the chemical structure of the channel should be known so that a physical theory for the permeation process could be constructed.

The biological channel that at present comes closest to meeting these criteria is the acetylcholine-activated channel found at a variety of vertebrate neuromuscular junctions and in the nervous system of certain invertebrates. For physiological studies, the frog neuromuscular junction is most frequently selected. In this preparation, the composition of the extracellular solution can be varied over quite a wide range, the main requirement being that the solution is approximately isoosmotic. Although the composition of the intracellular medium cannot be readily varied, the activities of the main permeant ionic species are known with reasonable precision. Reversal potentials are easily measured, and single-channel currents can be measured either directly (25) or estimated accurately but indirectly through fluctuation analysis (1). Because all the alkali metal and alkaline earth cations pass through this channel, the electrostatic interactions of ions with the channel structure can be studied by varying ion radius and valence (C. A. Lewis and C. F. Stevens, *unpublished observations*). The chemical structure of the channel is not yet known, but the acetylcholine receptor— and presumably the associated channel—have been isolated and purified and are currently being subjected to intensive biochemical study in a number of laboratories throughout the world (see articles in ref. 7); it seems likely that this channel will be the one whose detailed structure is first worked out.

Because the acetylcholine-activated channel [epc (end-plate current channel)] seems to present a number of advantages, we have embarked on a program whose ultimate goal is to account in physical terms for ion permeation through this particular channel. We hope, of course, that the epc channel will serve as a good model system in which to study ion permeation in general; that is, we would like to believe that what we learn in the epc channel is true for ion movements in other channel types, although the degree of generality of our results must be discovered by comparative studies.

We report here some of our preliminary studies on ion movements through epc channels. Briefly, we have found that the traditional electrodiffusion picture of ions moving independently of one another is inadequate because the various ions that can enter the channel behave as if they compete for a site that any ion must occupy before it can proceed through the channel. The simplest possible mechanism that embodies this type of competition—a channel with a single "site" surrounded by energy barriers—can account quantitatively for the experimental observations presented here.

EXPERIMENTAL METHODS

The experimental protocol involved making three different types of measurements in solutions with varying Na and divalent cation concentrations. The measurements were the following:

a) reversal potential, V_o
b) single-channel conductance, γ, at a fixed voltage, which was usually -70 mV
c) single-channel conductance as a function of holding potential over the range of -120 to -50 mV

Sucrose was used as the sodium substitute; consequently, the possible effects of the varying ionic strength in the different solutions was taken into account theoretically as described later. The sodium concentrations ranged from normal (115 mM) to 1 mM, 4 mM HEPES was the buffer used, and the pH of the solutions was 7.4. Details of the experimental procedures are described elsewhere (22).

For the theoretical calculations described below, a temperature of 15°C was used along with ion activities. The extracellular activities were calculated using data from Robinson and Stokes (27) and from Butler (6) and Shatkay (31). The intracellular activities of Na and K were measured using ion-selective electrodes, and the average values were 80 and 6.5 mM for K and Na, respectively.

The χ^2-test was used as an objective measure of the goodness of fit of the various theories to the experimental observations. The following formula was used to calculate χ^2:

$$\chi^2 = \sum_{i=1}^{n} \frac{(i_{exp} - i_{theory})^2}{i_{theory}}$$

where n is the number of comparisons, i_{exp} is the mean experimental value of γ or V_o, and i_{theory} is the value predicted by the different theories. The degrees of freedom (d.f.) are the total number of comparisons minus the number of theoretical parameters that were estimated from the set of experimental values. The probability of obtaining by chance a particular value for χ^2 or a lower value is indicated by P, so that a small value for P indicates that the theory provides a good fit to the experimental data.

EXPERIMENTAL OBSERVATIONS

The reversal potential varied, as shown in Fig. 1, approximately logarithmically with sodium concentration for concentrations greater than about 10% of normal and seemed to approach a limiting value close to −60 mV for very low sodium concentrations (note that the external solution always contained the following concentrations—2 mM Ca and 2.5 mM K). Isotonic Ca gave a reversal potential about like that of a solution containing 30% normal sodium; the Ca activity is approximately equal to the sodium activity in this circumstance.

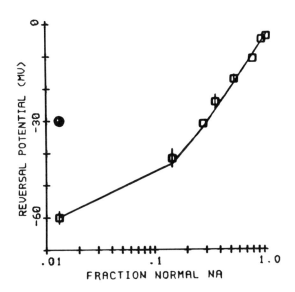

FIG. 1. Reversal potential as a function of external sodium ion activity expressed as a fraction of the normal $[Na]_o$. The experiments were performed at temperatures ranging from 10 to 15°C. The squares are experimental data for solutions containing 2 mM Ca, whereas the circle is the experimental point for a solution with 80 mM calcium Ringer's along with 1 mM Na. The SEM is indicated if it is larger than the square. The smoothcurve and cross (inside circle) are theoretical predictions for a symmetrical two-barrier theory with $\sigma = 0.01$ electronic charges/A^2. (The lines in the figure have no theoretical significance and are used for clarity to connect the theoretical values corresponding to the indicated experimental points.)

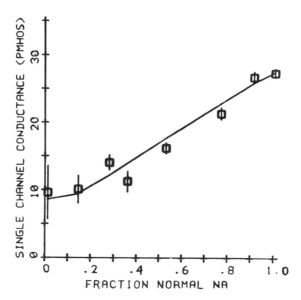

FIG. 2. Single-channel conductance as a function of external sodium ion activities expressed as a fraction of the normal $[Na]_o$. The experiments were performed at temperatures ranging from 10 to 15°C, and the theoretical calculations are for a temperature of 15°C. The squares indicate experimental data for solutions containing 2 mm Ca, and the SEM is indicated if it is larger than the square. The curve indicates a fit of the two-barrier theory for a symmetrical channel with $\sigma = 0.01$ electronic charges/A², $\delta = 0.5$, and $\alpha_2 = \alpha_3 = 1$. This fit predicts that γ for Ca Ringer's should be 9.3 pS (compared with the observed value of 10.1 $\pm$ 0.4 pS) and γ for 75% Na with 10 mm Ca should be 14.0 pS (compared with the observed value of 14.7 + 1.2 pS). (The lines in the figure have no theoretical significance and are used for clarity to connect the theoretical values corresponding to the indicated experimental points.)

The single-channel conductance varied approximately linearly with extracellular sodium concentration for concentrations greater than about 10% of normal and appeared to approach a limiting value of about 10 pS for very low sodium concentrations (see Fig. 2). The isotonic Ca solution gave about the same single-channel conductance as a sodium solution with about 1% of normal sodium. A solution containing 75% normal sodium (with 2 mm Ca) gave a single-channel conductance of a little over 20 pS, and increasing the Ca concentration of this solution to 10 mm caused the single-channel conductance to decrease about 25%.

The single-channel conductance γ does not vary strongly with voltage over the range of membrane potentials from -120 to -50 mV and for concentrations between 25 and 100% of normal extracellular sodium. The regression line relating conductance and voltage had a slope of -0.73 ± 0.039 pS/mV.

TRADITIONAL THEORIES: THE APPROACHES OF TAKEUCHI AND OF GOLDMAN, HODGKIN, AND KATZ

The theories that have traditionally been applied to ion movements through epc channels are that of Takeuchi (32–34) and of Goldman, Hodgkin, and

Katz (13,19). These two different approaches share a central assumption: in both cases ions are supposed to move through the channels independently so that the probability of an ion traversing the channel is not influenced by any of the other ions present. This is the only assumption that needs to be made for the Takeuchi approach, whereas the theory used by Goldman and by Hodgkin and Katz also required, in its original form, that the electric field inside the membrane be constant; however, it is possible to derive an equation similar to their reversal potential equation using approaches that relax this requirement (4,8,28,35).

We wish now to test these two approaches by comparing the predictions they make with the data presented in Figs. 1 and 2. The first step in this program is to present the theories in a form suitable for such a test.

According to the Takeuchi approach, the reversal potential is a weighted average of the Nernst equilibrium potentials for the different permeant ions, with the weighting factor equal to a conductance ratio g_x/g_K where g_x is the conductance of ion x and is defined by $I_x = g_x (V - V_x)$ with V equal to the holding potential, V_x equal to the Nernst equilibrium potential for ion x, and I_x the current carried by ion x. The Nernst potential V_x for ion x is given by

$$V_x = \frac{RT}{Z_x F} ln \frac{[X]_o}{[X]_i}$$

where R is the gas constant, T the temperature in °K, F the Faraday, Z_x the valence of ion x, and $[X]_o$ is the extracellular and $[X]_i$ the intracellular activities of x. The following equation gives the reversal potential for a channel system through which Na, K, and Ca ions move—these three ions have been shown to be permeant at epc channels (32,34).

$$V_o = \frac{g_K V_K + g_{Na} V_{Na} + g_{Ca} V_{Ca}}{g_K + g_{Na} + g_{Ca}} \qquad [1]$$

The advantage of the Takeuchi approach is that, in general, it only needs to assume ion flux independence, whereas its weakness is that the functional relationship between conductance ratios and external activities is not known. In practice, the additional assumption is usually made that the conductance ratios remain constant and are independent of ion concentrations. The following equation is derived elsewhere (22) and uses single-channel conductance measurements in different solutions as an empirical indication of the functional relationship between conductance ratios and external concentrations:

$$V_o = \frac{V_K + \dfrac{\gamma_{meas} - (\gamma_K + \gamma_{Ca})}{\gamma_K} V_{Na} + \dfrac{\gamma_{Ca}}{\gamma_K} V_{Ca}}{1 + \dfrac{\gamma_{meas} - (\gamma_K + \gamma_{Ca})}{\gamma_K} + \dfrac{\gamma_{Ca}}{\gamma_K}} \qquad [2]$$

where γ_{meas} is the experimentally observed single-channel conductance and $\gamma_K + \gamma_{Ca}$ are the K$^+$ and Ca^{2+} contributions, respectively, to the observed single-channel conductance.

By comparing the predictions of Eq. [2] with the data in Fig. 1, we can test the adequacy of the Takeuchi approach for epc channels. The Nernst potentials V_{Na}, V_K, and V_{Ca} are calculated from the known extracellular and measured intracellular ion activities, and the constant $\gamma_K + \gamma_{Ca}$ is estimated from the data presented in Fig. 2. If the Takeuchi approach is adequate, then Eq. [2] should provide a good fit to the Fig. 1 reversal potential data with benefit of only a single adjustable parameter, γ_{Ca}/γ_K.

In constant field theory, the laws of electrodiffusion are used to predict how currents and conductances depend on concentration. Again, this theory makes the assumption of ion flux independence. As originally derived by Goldman and Hodgkin and Katz, their reversal potential equation is only for monovalent cations. Modifying the equation to include a divalent cation, Ca^{2+}, one introduces mathematical complications. One way of expressing the result is the following (see ref. 22, for a derivation; see ref. 20 for an alternate approach):

$$V_o = \frac{RT}{F} \ln \frac{[Na]_o + P_K/P_{Na}[K]_o + 4P'[Ca]_o}{[Na]_i + P_K/P_{Na}[K]_i + 4P'[Ca]_i e^{FVo/RT}} \qquad [3]$$

where

$$P' = \frac{P_{Ca}}{P_{Na}} \left(\frac{1}{1 + e^{FVo/RT}} \right)$$

The current carried by ion x is given by the following equation:

$$I_x = \frac{Z_x^2 F^2}{RT} P_x V \left(\frac{[X]_o - [X]_i e^{Z_x FV/RT}}{1 - e^{Z_x FV/RT}} \right) \qquad [4]$$

The chord conductance is $G = I/V - V_o$, and this is the calculated parameter to be compared with the experimentally observed single-channel conductance measurements. The idea here is to estimate, through Eq. [3], the permeability ratios P_K/P_{Na} and P_{Ca}/P_{Na} from the reversal potential data in Fig. 1, and then use Eq. [4] to predict the single-channel conductance data of Fig. 2. If this approach is adequate, single-channel conductance should be accurately predicted with only a single adjustable parameter—the permeability coefficient P_{Na}.

The constant field theory can be modified to include ion interactions through surface charge effects (10). If a fixed, negative surface charge density, σ, exists on the outer surface of the membrane, then cations will tend to be concentrated on the surface of the membrane according to the Boltzmann relationship

$$[X]_s = [X]_o e^{-Z_x F\psi/RT}$$

where $[X]_s$ is the surface concentration of ion x, $[X]_o$ is the bulk concentration, and ψ is the surface potential and is related to extracellular ion concentrations and the value for σ according to the following formula (see refs. 18 and 24 for underlying assumptions):

$$\sigma = \frac{1}{272} \left[\sum_x [X]_o \left(\exp \frac{-Z_x F\psi}{RT} - 1 \right) \right]^{1/2} \qquad [5]$$

where σ is the surface charge density in electron charges/A². Equations [3] and [4] are modified in the following way to include the effects of surface charge; V and V_o are replaced by the new potential drop across the membrane of V-ψ and V_o-ψ, whereas the extracellular activities are replaced with surface activities as given by the Boltzmann relationship.

Surface charge effects can lead to ion interactions. The surface potential, ψ, depends on the concentrations of all of the ions present in the external solutions as given by Eq. [5]. A consequence of this is the fact that increasing the divalent cation concentration results in decreasing the negativity of the surface potential due to more screening of fixed negative charges. The fact that ψ is less negative results in the reduction somewhat of the surface concentration of Na so that the overall Na contribution to the single-channel conductance is decreased.

TESTS FOR THEORIES OF ION PERMEATION

Various theories were tested for their ability to predict the three types of experimental observations made in the different solutions. In particular, four different theories were tested: (a) the Takeuchi approach, (b) constant field theory with surface charge, (c) a theory in which it was assumed that ions jumped independently over a single energy barrier to pass through the membrane, and (d) a two-barrier–one-well theory that includes ion competition for occupancy of an energy well reached by jumping over energy barriers. The results of fitting the first three theories have been previously reported (22) and are briefly summarized below. The results of fitting the fourth theory to experimental observations are reported here in detail.

The general result that was previously reported is that none of the first three theories is completely adequate in describing all of the experimental observations. From the failure of these theories, certain conclusions can be drawn.

The first conclusion reached from the preceding study is that ion interactions occur at epc channels. This is a direct result of the fact that the modified Takeuchi approach, Eq. [2], is completely inadequate for describing the experimental results. Specifically, when γ_K and γ_{Ca} in Eq. [2] are assumed to be independent of the extracellular sodium concentration, their sum can be estimated from the limiting value of γ as $[Na]_o$ approaches 0. Then Eq. [2] must fit the reversal potential data of Fig. 1 using the observed γ values (Fig. 2) and a single adjustable parameter γ_{Ca}/γ_K. When this procedure is carried out, Eq. [2] is found not to provide satisfactory predictions of reversal potentials for the full range of extracellular sodium concentrations. For example, if γ_{Ca} and γ_K are selected so that V_o is fitted by Eq. [2] for normal sodium concentration, the predictions for the concentrations below 50% normal are in error by more than 20 mV. The only assumption in the derivation of this equation that has no experimental support whatever is the assumption of ion flux independence; therefore, the conclusion is reached that this assumption is invalid.

The ion interactions occurring at epc channels could involve surface charge effects or competition for a binding site, or both. The possibility that surface charge effects alone could account for the observed ion interactions was tested using two different theories—constant field theory and a one-barrier theory—both with a fixed negative surface charge density on the outer surface of the membrane. It was not, unfortunately, possible to modify the Takeuchi approach to include surface charge because the dependence of γ_{Na}, γ_K, and γ_{Ca} on surface concentrations of the relevant ions is unknown. The two theories used allow ion interactions in the form of surface charge effects to occur at the channel, but still assume ions move independently of one another. The result was that both theories could adequately account for all of the reversal potential measurements, but could not adequately predict the single-channel conductances under all of the experimental conditions. In conclusion, then, postulating the existence of surface charge effects is sufficient to account for V_o measurements, but the existence of additional interactions needs to be postulated in order to account for all of the γ values. As we show below, additional interactions in the form of competition for a binding site is sufficient to predict all of the experimental observations, including the single-channel conductance measurements.

Ion permeation through membrane channels has recently been viewed as similar to diffusion through a one-dimensional crystal (3,9,12,14–17,21,26,29, 36,37). For such theories, ions could interact through competition for an internal membrane binding site. The exact energy profile experienced by an ion traversing the epc channel must be hopelessly complex. Because jump rates depend exponentially on barrier heights, however, relatively small differences in barrier heights cause large differences in jump rates so that only one or a few steps can be expected to be rate limiting. One might anticipate, then, that as a consequence of the fact that the largest energy barriers dominate ion motion in the channel, a limiting approximation might be likely to incorporate the main features of ion permeation.

The simplest theory that incorporates competition for a site is one in which there is a single binding site with an energy barrier on each side. Not only is this the simplest theory, but this two-barrier theory might be anticipated to be a reasonable approximation to actual physical reality in which only the largest barriers governing access to the site are included. A representation of this model is shown in Fig. 3.

In solving the two-barrier theory for the predicted current and reversal potential, the assumption is made that the energy barriers are different for each type of ion and that these barrier heights are not changed by the presence of the other ions in the bathing solution. Furthermore, the assumption is also made that the binding site must be empty before it can be occupied by an ion. This is a limiting case of a more general knock-on type of model, as described

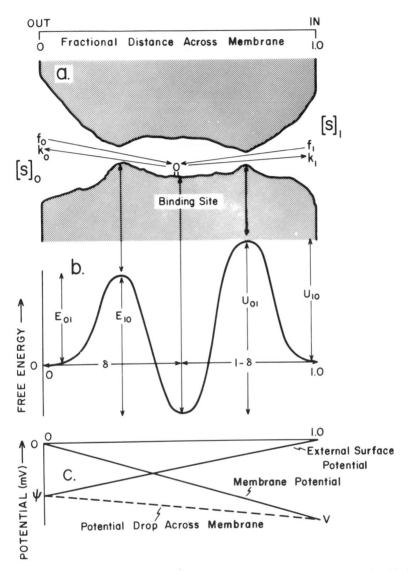

FIG. 3. The two-barrier model. **a:** A hypothetical channel with two barriers and a binding site. The rates of binding are given by f_0 and f_1, whereas k_0 and k_1 indicate rates of unbinding. **b:** The energy profile for the hypothetical channel with the energy well located at δ. **c:** The voltage profile for the hypothetical channel with the true potential drop across the membrane being given by V-Ψ.

by Armstrong (2), in which the probability of one ion displacing another ion from the binding site is essentially zero.

The current equation (derived in the Appendix) for the specific case of three ions with subscripts Na = 1, K = 2, and Ca = 3 is given by the following equation:[1]

$$I = FP_0[e^{\frac{-\delta FV}{RT}}(k'_{01}b_1 + k'_{02}b_2 - f'_{01}C_{01} - f'_{02}C_{02})$$

$$+ e^{-\delta FV/RT}(2k'_{03}b_3 - 2f'_{03}C_{03})] \quad [6]$$

with

$$b_1 = \frac{f'_{01}C_{01} + f'_{11}e^{FV/2RT}C_{11}}{k'_{01} + k'_{11}e^{-FV/2RT}}$$

$$b_2 = \frac{f'_{02}C_{02} + f'_{12}e^{FV/2RT}C_{12}}{k'_{02} + k'_{12}e^{-FV/2RT}}$$

$$b_3 = \frac{f'_{03}C_{03} + f'_{13}C_{13}e^{FV/RT}}{k'_{03} + k'_{13}e^{-FV/RT}}$$

$$P_0 = \frac{1}{1 + e^{-\delta FV/RT}(b_1 + b_2 + e^{\delta FV/RT}b_3)}$$

where P_0 is the probability that the binding site is empty; C_{01}, C_{02}, and C_{03} refer to the external activities of Na, K, and Ca, respectively; C_{11}, C_{12}, and C_{13} refer to the intracellular activities of Na, K, and Ca, respectively; and f'_{ij} and k'_{ij} are the voltage-independent part of the rate constants as defined in Fig. 3. The chord conductance is given simply by $\gamma = I/V - V_o$.

The calculation of the reversal potential turns out to be complicated considerably by the inclusion of a divalent cation. Because an explicit expression for the reversal potential could not be obtained, the following is one way of presenting the result (see Appendix):

$$V_o = \frac{RT}{F}\ln\frac{C_{01} + \frac{f'_{02}}{f'_{01}}C_{02}\left[1 + \left(\frac{\alpha_2 - 1}{\beta}\right)A\right] + 2\frac{f'_{03}}{f'_{01}}C_{03}\left[1 + \frac{M}{B}\right]}{C_{11} + \frac{f'_{02}}{f'_{01}}C_{12} + 2\frac{f'_{03}}{f'_{01}}C_{13}} \quad [7]$$

with

$$\alpha_2 = \frac{k'_{01}\,k'_{12}}{k'_{11}\,k'_{02}} \qquad \alpha_3 = \frac{k'_{01}\,k'_{13}}{k'_{11}\,k'_{03}}$$

---

[1] For the rate constants f_{ij} and k_{ij} the subscript i indicates the particular side of the membrane with 0 = outside and 1 = inside, whereas j indicates the ion species.

$$A = \left(e^{-3 FV_0/2 RT} - \frac{C_{12}}{C_{02}} e^{-FV_0/2 RT} \right) \left(e^{FV_0/RT} + \frac{k'_{13}}{k'_{03}} \right)$$

$$B = \beta e^{\delta FV_0/2 RT}$$

$$\beta = e^{-FV_0/2 RT} + \frac{k'_{12}}{k'_{02}} e^{-FV_0/RT} + \frac{k'_{13}}{k'_{03}} e^{-3 FV_0/2 RT} + \frac{k'_{12}}{k'_{02}} \frac{k'_{13}}{k'_{03}} e^{-2 FV_0/RT}$$

$$M = \frac{C_{13}}{C_{03}} e^{F(\delta+1) V_0/2 RT} - \frac{C_{13}}{C_{03}} e^{FV_0/2 RT} \left(\frac{k'_{13}}{k'_{03}} + \frac{k'_{12}}{k'_{02}} \alpha_3 \right)$$

$$- \frac{C_{13}}{C_{03}} e^{F\delta V_0/2 RT} \left(\alpha_3 - \frac{k'_{12}}{k'_{02}} \right)$$

$$- \frac{C_{13}}{C_{03}} \frac{k'_{12}}{k'_{02}} \frac{k'_{13}}{k'_{03}} + e^{F(\delta-1) V_0/2 RT} \left(\frac{C_{13}}{C_{03}} \frac{k'_{13}}{k'_{03}} - 1 \right)$$

$$+ e^{F(\delta-2) V_0/2 RT} \left(\frac{C_{13}}{C_{03}} \frac{k'_{12}}{k'_{02}} \frac{k'_{13}}{k'_{03}} - \frac{k'_{12}}{k'_{02}} \right)$$

$$+ \alpha_3 e^{-FV_0/RT} - \frac{k'_{13}}{k'_{03}} e^{F(\delta-3) V_0/2 RT} + e^{-3 FV_0/2 RT} \frac{k'_{13}}{k'_{03}} (1 + \alpha_3)$$

$$+ \frac{k'_{12}}{k'_{02}} \frac{k'_{13}}{k'_{03}} e^{-2 FV_0/RT} - \frac{k'_{12}}{k'_{02}} \frac{k'_{13}}{k'_{03}} e^{F(\delta-4) V_0/2 RT}$$

Equation [7] looks quite formidable, and not all of the parameters involved have obvious physical significance. α_2 and α_3 are meaningful parameters—a measure of how well the "constant offset energy peak condition" is met (see ref. 16 for a further discussion). If the two barrier heights for potassium (or calcium) are higher or lower than the corresponding ones for sodium by a constant amount, then $\alpha_2 = 1$ (or $\alpha_3 = 1$), and the constant offset energy peak condition is said to be met for that ion. There is no *a priori* reason for assuming that the constant offset energy peak condition either should be met or should not be met for epc channels. f'_{02}/f'_{01} and f'_{03}/f'_{01} are the ratios of the outer barrier heights relative to those of sodium and could be considered a sort of $(P_K/P_{Na})_{effective}$ and $(P_{Ca}/P_{Na})_{effective}$. If $\alpha_2 = 1$, then f'_{02}/f'_{01} can be directly associated with the P_K/P_{Na} in the constant field voltage equation (Eq. [3]). Also, k'_{13}/k'_{03}, k'_{12}/k'_{02}, and k'_{11}/k'_{01} indicate the degree of asymmetry in the membrane and equal 1 for the completely symmetrical membrane with equal barrier heights for an ion species.

There are certain limiting cases in which Eq. [7] reduces to a recognizable result. If the constant offset energy peak condition is met (i.e., $\alpha_2 = \alpha_3 = 1$) and in the limit of $FV_0/RT \to 0$, then Eq. [7] reduces to the following equation, which is of the same form as the one-barrier and constant field result in the same limit:

$$V_o = \frac{RT}{F} \ln \frac{C_{01} + \frac{f'_{02}}{f'_{01}} C_{02} + \frac{2f'_{03}}{f'_{01}} C_{03}}{C_{11} + \frac{f'_{02}}{f'_{01}} C_{12} + \frac{2f'_{03}}{f'_{01}} C_{13}}$$

Another limiting case is when calcium is the only ion present. In this case Eq. [7] reduces to $V_o = (RT/2F)ln\ ([Ca]_o/[Ca]_i)$ which is the Nernst equilibrium potential for calcium.

As written, Eqs. [6] and [7] allow ion interactions only to occur through competition for a binding site. However, the possibility exists that some of the observed interactions could be due to surface charge effects; therefore, the equations need to be modified to include this possibility. Only slight modifications are needed—V is replaced every place it occurs by the new potential drop across the membrane $(V - \Psi)$, V_o is replaced by $(V_o - \Psi)$, and the relevant external activities to be used are surface activities.

Because the equations resulting from even this simplest theory—one site and two barriers per ion—are quite complex, three representative limiting cases were selected for detailed study. One limiting case was a completely symmetrical membrane (i.e., each ion sees inner and outer barriers of the same height, but the heights are different for different species) with the constant offset energy peak condition being met (i.e., $\alpha_2 = \alpha_3 = 1$). As soon as one energy barrier is larger than the other, the result is a great difference in jump rates due to the exponential dependence on energy differences. The other two limiting cases, then, were asymmetrical membranes with either the outer or inner barrier being rate limiting.

The mean experimental V_o values in Fig. 1 were fitted using Eq. [7]. The location of the well was arbitrarily placed at the middle of the potential drop across the membrane (i.e., $\delta = 0.5$). Values for the fixed negative surface charge density of 0.005 to 0.01 electronic charges/A^2 were selected because these appeared to be physiological values for other biological systems (5,11,18,24,30). This leaves five parameters to be estimated from Eq. [7] in order to fit the V_o values in Fig. 1—the two permeability ratios P_K/P_{Na}, P_{Ca}/P_{Na}, one of the ratios such as k'_{13}/k'_{03} that indicates the degree of asymmetry of the channel, and the two parameters α_2 and α_3 that indicate if the constant offset energy peak condition is met for K or Ca, or both. In practice, limiting cases were studied such that $\alpha_2 = \alpha_3 = 1$ and $k'_{13}/k'_{03} = 1$, 0, or ≥ 0 for a symmetrical channel, one with the inner barrier much larger or a channel with the outer barrier much larger, respectively. For these limiting cases, then, there are actually only two parameters to be estimated from Fig. 1—the two permeability ratios.

Fitting eq. [6] to the data in Fig. 2 requires four additional parameters to be estimated. These are the three well depths for Na, K, and Ca and the absolute barrier height for the reference ion, Na.

One result from the study of these three limiting cases was that it is impossible to fit the reversal potential data adequately if the surface charge density is assumed to be 0 (e.g., for $\alpha_2 = \alpha_3 = 1$ and $\delta = 0.5$, $\chi^2 = 4.6$, d.f. $= 8$, $P \leq 0.25$). The lack of fit with $\sigma = 0$ implies that competition for a binding site cannot account for all of the ion interactions. It seems most likely, then, that ion interactions occur at the epc channel in the form of both surface charge effects and binding site competition.

All three limiting cases were able to fit adequately the reversal potential data if $\sigma > 0.001$ electronic charges/A^2. These fits are indistinguishable from the curve in Fig. 1, which is for $\sigma = 0.01$ electronic charges/A^2. However, only the symmetrical channel and the asymmetrical one with the inner barrier being the larger were able to fit satisfactorily the single-channel conductance measurements. For example, a fit for a symmetrical channel with $\sigma = 0.01$ electronic charge/A^2 is shown in Fig. 2. (Ψ for normal Ringer's $= -101$mV) ($\chi^2 = 1.5$, d.f. $= 7$, $P \leq 0.025$). For this fit, $P_K/P_{Na} = 1.2$ and $P_{Ca}/P_{Na} = 0.10$, the barrier heights for Na, K, and Ca are 5.2, 5.1, and 6.6 kcal/mole, respectively, and K and Ca both see wells of about -3 kcal/mole, whereas Na doesn't see a well.

The energy profiles for the best fits for the two limiting cases with the inner barrier larger ($\sigma = 0.005$ electronic charges/A^2, $\delta = 0.5$) and two equal barriers ($\delta = 0.01$ electronic charges/A^2, $\delta = 0.5$) are shown in Fig. 4 and have some features in common. For example, the barrier heights for the rate-limiting barrier is the same in both cases as are the well depths. The Ca well needs to be about -3 kcal/mole in order to predict the correct voltage dependence of γ. It is possible to satisfy all of the observed data without postulating a well for Na.

The two limiting cases also give similar results for the effective permeability ratios. $(P_K/P_{Na})_{effective} = 1.2$, whereas the values for $(P_{Ca}/P_{Na})_{effective}$ depend on the assumed surface charge density and are relatively independent of δ and k_{13}'/k_{03}'.

In conclusion, the two-barrier theory can adequately account for all of the experimental observations, but in general does not provide a unique set of values for all of the parameters. The general two-barrier model can be simplified if the constant offset energy peak condition is met (i.e., $\alpha_2 = \alpha_3 = 1$). In this situation $(P_K/P_{Na})_{effective}$ always equals approximately 1.2, whereas $(P_{Ca}/P_{Na})_{effective}$ ranges from 0.05 to 0.2 depending on the assumed surface charge density. The energy barrier for Na$^+$ is slightly larger than that for K$^+$, whereas the energy barrier height for C^{2+} is 1 to 2 kcal/mole higher. There may or may not be wells present for Na$^+$, but Ca^{2+} must see an energy well of about -3 kcal/mole.

In summary, ion interactions occur at epc channels, and these probably involve both surface charge effects and competition for a binding site. The traditional theories using the approaches of the Takeuchis and of Goldman, Hodgkin, and Katz are unable to predict simultaneously the single-channel conductances and reversal potentials for a range of extracellular sodium activities. A simple theory with two barriers and a well is able to account for all of the experimental

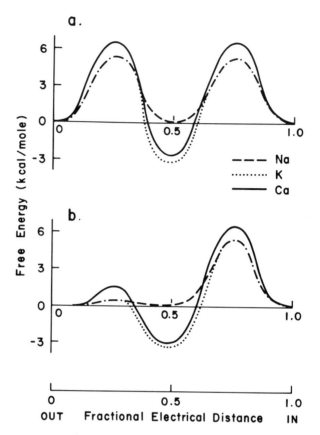

FIG. 4. Two energy profiles for the epc channel for the two-barrier theory that satisfactorily predicts all of the observed experimental data. The calculations are for a temperature of 15°C. **a:** A symmetrical channel with constant offset energy peak conditions met (i.e., $\alpha_2 = \alpha_3 = 1$) and with $\sigma = 0.01$ electronic charges/A^2 and $\delta = 0.5$. **b:** An asymmetrical channel with the inner barrier rate limiting, $\sigma = 0.005$ electronic charges/A^2 (also $\alpha_2 = \alpha_3 = 1$ and $\delta = 0.5$).

observations. A challenge for the future then is to devise a way to further refine this picture and to construct a physical theory to account for these barrier heights and well depths.

APPENDIX

Derivation of Equations for Two-Barrier Theory

In this model, f_{ix} refers to rates of binding, k_{ix} refers to rates of unbinding, and both are expressed in units of sec^{-1} (see footnote[2] below). It is convenient to separate

---

[2] The subscript i refers to the side of the membrane with $0 =$ outside and $1 =$ inside the membrane. The subscript x refers to the particular ion species with $Na = 1$, $K = 2$, and $Ca = 3$.

the rate constants into a voltage term multiplied by a voltage-independent constant denoted by a prime ('). The four rate constants for ion x are given by Eyring rate theory as the following:

$$k_{0x} = v_1 e^{-E_{10}/RT} e^{Z_x \delta FV/2RT} = k'_{0x} e^{Z_x \delta FV/2RT} \qquad \text{[A–1]}$$

$$k_{1x} = v_1 e^{-U_{01}/RT} e^{-Z_x(1-\delta)FV/2RT} = k'_{1x} e^{-Z_x(1-\delta)FV/2RT} \qquad \text{[A–2]}$$

$$f_{0x} = [X]_0 v_2 e^{-E_{01}/RT} e^{-Z_x \delta FV/2RT} = f'_{0x}[X]_0 e^{-Z_x \delta FV/2RT} \qquad \text{[A–3]}$$

$$f_{1x} = [X]_1 v_2 e^{-U_{10}/RT} e^{Z_x(1-\delta)FV/2RT} = f'_{1x}[X]_1 e^{Z_x(1-\delta)FV/2RT} \qquad \text{[A–4]}$$

where v_1 and v_2 are vibration frequencies and were assumed to be 10^{13}/sec and 10^{11}/molar-sec, respectively.[3] The energies E_{10}, U_{01}, E_{01}, and U_{10} are indicated in Fig. 3b.

The binding site can be in one of four states—either empty or occupied by ion 1, 2, or 3. The probabilities of being in these states are denoted by P_0, P_1, P_2, and P_3, respectively.

The assumption is made that the binding site has to be empty before an ion can bind; therefore, the flux of ion x going to the binding site is $P_0(f_{1x} + f_{0x})$. Similarly, the site must be occupied by ion x before unbinding can occur; therefore, the flux of ion x leaving the binding site is given by $P_x(k_{0x} + k_{1x})$. In the steady state, the flux of ion x entering the binding site equals the flux of ion x leaving it. Therefore, the following equations hold in the steady state for the epc channel:

$$P_0(f_{11} + f_{01}) = P_1(k_{01} + k_{11}) \qquad \text{[A–5]}$$

$$P_0(f_{12} + f_{02}) = P_2(k_{02} + k_{12}) \qquad \text{[A–6]}$$

$$P_0(f_{13}) + f_{03}) = P_3(k_{03} + k_{13}) \qquad \text{[A–7]}$$

Using the constraint that the sum of the probabilities must equal 1 gives the following additional equation:

$$P_0 + P_1 + P_2 + P_3 = 1 \qquad \text{[A–8]}$$

P_1, P_2, and P_3 can be written in terms of P_0 using Eqs. [A–5] to [A–7] above giving the following result:

$$P_x = P_0 \frac{f_{1x} + f_{0x}}{k_{0x} + k_{1x}} \qquad \text{[A–9]}$$

Substituting these results into Eq. [A–8] gives the following expression for P_0:

$$P_0 = \frac{1}{1 + \dfrac{f_{01} + f_{11}}{k_{01} + k_{11}} + \dfrac{f_{02} + f_{12}}{k_{02} + k_{12}} + \dfrac{f_{03} + f_{13}}{k_{03} + k_{13}}} \qquad \text{[A–10]}$$

The net flux of ion x is the difference in the one-way fluxes $P_x k_{0x} - P_0 f_{0x}$ which gives $I_x = Z_x F(P_x k_{0x} - P_0 f_{0x})$ for the current carried by ion x. The general current equation, then, for n ions permeating a channel is given by

$$I = F\left[\sum_{x=1}^{n} Z_x k_{0x} P_0 \frac{f_{1x} + f_{0x}}{k_{0x} + k_{1x}} - P_0 \sum_{x=1}^{n} Z_x f_{0x} \right] \qquad \text{[A–11]}$$

If Eq. [A–11] is written out in full for Na, K, and Ca, then the following equation is obtained for the total current through the channel:

---

[3] $v_1 = 10^{13}$/sec is an average vibration frequency for solids, and $v_2 = 10^{11}$/molar-sec is an average vibration frequency for bimolecular reactions calculated from simple collision theory. See ref. 23.

$$I = FP_0[e^{-\delta FV/2RT}(k_{01}'b_1 + k_{02}'b_2 - f_{01}'C_{01} - f_{02}'C_{02})$$

$$+ e^{-\delta FV/RT}(2k_{03}'b_3 - 2f_{03}'C_{03})] \qquad [A\text{--}12]$$

with

$$b_1 = \frac{f_{01}'C_{01} + f_{11}'e^{FV/2RT}C_{11}}{k_{01}' + k_{11}'e^{-FV/2RT}}$$

$$b_2 = \frac{f_{02}'C_{02} + f_{12}'e^{FV/2RT}C_{12}}{k_{02}' + k_{12}'e^{-FV/2RT}}$$

$$b_3 = \frac{f_{03}'C_{03} + f_{13}'C_{13}e^{FV/RT}}{k_{03}' + k_{13}'e^{-FV/RT}}$$

$$P_0 = \frac{1}{1 + e^{-\delta FV/RT}(b_1 + b_2 + e^{-\delta FV/RT}b_3)}$$

The procedure involved in finding the reversal potential is to set the current equation equal to 0 and solve for V_0. Dividing through by $f_{01}'k_{11}'k_{02}'$ simplifies the expression.

It is easier to solve for the reversal potential if the constant offset energy peak condition is met (i.e., $\alpha_2 = \alpha_3 = 1$). If this is the case, then an equation of the following general form is obtained when the current equation is set equal to 0:

$$A \sum_x Z_x \frac{P_x}{P_1} C_{0x} = B \sum_x Z_x \frac{P_x}{P_1} C_{1x} \qquad [A\text{--}13]$$

where x is summed over all of the ions present and the P_x now indicate permeability coefficients in terms of ratios of barrier heights.[4] A and B are exponential terms of the following form:

$$A = \sum_j a_j e^{b_j V_0}; \quad B = \sum_j C_j e^{d_j V_0}$$

where j is the total number of exponential terms and is equal to 4 for three ions. A and B are related to each other in the following way:

$$A e^{FV_0/RT} = B$$

Substituting this relationship into Eq. [A–13] gives the following:

$$e^{FV_0/RT} = \frac{\sum_x Z_x \frac{P_x}{P_1} C_{0x}}{\sum_x Z_x \frac{P_x}{P_1} C_{1x}}$$

which immediately gives the Goldman–Hodgkin–Katz voltage equation for monovalent cations.

If the constant offset energy peak condition is not met, then the algebra is messier. The strategy involved is to force the general equation that results when the current (e.g. Eq. [A–12]) is set equal to 0 into a form that looks like Eq. [A–13] so that the same technique can be used to solve for V_0. Now, the general equation can be written in the following form:

$$A \sum_x Z_x \frac{P_x}{P_1} C_{0x} + M_1 = B \sum_x Z_x \frac{P_x}{P_1} C_{1x} + M_2$$

---

[4] $P_x \equiv \dfrac{e^{-E_{0x}/RT}}{e^{-E_{01}/RT}}$

where A and B are related by the same relationship. Solving for $e^{FV_o/RT}$ now gives the following equation:

$$e^{FV_o/RT} = \frac{\sum_x Z_x \dfrac{P_x}{P_1} C_{ox} + \dfrac{M_1 - M_2}{A}}{\sum_x Z_x \dfrac{P_x}{P_1} C_{1x}}$$

Evaluating M_1, M_2, and A, collecting terms, and taking natural logarithms of both sides gives the following equation for the reversal potential:

$$V_o = \frac{RT}{F} \ln \frac{C_{01} + \dfrac{f'_{02}}{f'_{01}} C_{02} \left[1 + \left(\dfrac{\alpha_2 - 1}{\beta}\right) A\right] + 2 \dfrac{f'_{03}}{f'_{01}} C'_{03} \left[1 + \dfrac{M}{B}\right]}{C_{11} + \dfrac{f'_{02}}{f'_{01}} C_{12} + 2 \dfrac{f'_{03}}{f'_{01}} C_{13}} \qquad \text{[A-14]}$$

with

$$\alpha_2 = \frac{k'_{01}}{k'_{11}} \frac{k'_{12}}{k'_{02}}, \quad \alpha_3 = \frac{k'_{01}}{k'_{11}} \frac{k'_{13}}{k'_{03}}$$

$$A = \left(e^{-3FV_o/2RT} - \frac{C_{12}}{C_{02}} e^{-FV_o/2RT}\right)\left(e^{FV_o/RT} + \frac{k'_{13}}{k'_{03}}\right)$$

$$B = \beta e^{\delta FV_o/2RT}$$

$$\beta = e^{-FV_o/2RT} + \frac{k'_{12}}{k'_{02}} e^{-FV_o/RT} + \frac{k'_{13}}{k'_{03}} e^{-3FV_o/2RT} + \frac{k'_{12}}{k'_{02}} \frac{k'_{13}}{k'_{03}} e^{-2FV_o/RT}$$

$$M = \frac{C_{13}}{C_{03}} e^{F(\delta+1)V_o/2RT} - \frac{C_{13}}{C_{03}} e^{FV_o/2RT}\left(\frac{k'_{13}}{k'_{03}} + \frac{k'_{12}}{k'_{02}} \alpha_3\right) - \frac{C_{13}}{C_{03}} e^{F\delta V_o/2RT}\left(\alpha_3 - \frac{k'_{12}}{k'_{02}}\right)$$

$$- \frac{C_{13}}{C_{03}} \frac{k'_{12}}{k'_{02}} \frac{k'_{13}}{k'_{03}} + e^{F(\delta-1)V_o/2RT}\left(\frac{C_{13}}{C_{03}} \frac{k'_{13}}{k'_{03}} - 1\right) + e^{F(\delta-2)V_o/2RT}\left(\frac{C_{13}}{C_{03}} \frac{k'_{12}}{k_{02}} \frac{k'_{13}}{k'_{03}} - \frac{k'_{12}}{k'_{02}}\right)$$

$$+ \alpha_3 e^{-FV_o/RT} - \frac{k'_{13}}{k'_{03}} e^{F(\delta-3)V_o/2RT} + e^{-3FV_o/2RT} \frac{k'_{13}}{k'_{03}} (1 + \alpha_3)$$

$$+ \frac{k'_{12}}{k'_{02}} \frac{k'_{13}}{k'_{03}} e^{-2FV_o/RT} - \frac{k'_{12}}{k'_{02}} \frac{k'_{13}}{k'_{03}} e^{F(\delta-4)V_o/2RT}$$

REFERENCES

1. Anderson, C. R., and Stevens, C. F. (1973): Voltage clamp analysis of acetylcholine produced end-plate current fluctuations at frog neuromuscular junction. *J. Physiol. (Lond.),* 235:655–691.
2. Armstrong, C. M. (1975): Potassium pores of nerve and muscle membranes. In: *Membranes,* Vol. 3, edited by G. Eisenman. Dekker, New York.
3. Baker, M. B. (1971): Ion Transport Through Nerve Membranes. Doctoral Dissertation. University of Washington, Seattle.
4. Barr, L. (1965): Membrane potential profiles and the Goldman equation. *J. Theor. Biol.,* 9:351–356.
5. Begenisich, T. (1975): Magnitude and location of surface charges on *Myxicola* giant axons. *J. Gen. Physiol.,* 66:47–65.
6. Butler, J. N. (1968): The thermodynamic activity of calcium ion in sodium chloride-calcium chloride electrolytes. *Biophys. J.,* 8:1426–1433.

7. *Cold Spring Harbor Symposia on Quantitative Biology, Vol. XL, The Synapse* (1976).
8. Conti, F., and Eisenman, G. (1965): The non-steady state membrane potential of ion exchangers with fixed sites. *Biophys. J.,* 5:247–256.
9. Eyring, H., Lumry, R., and Woodbury, J. W. (1949): Some applications of modern rate theory to physiological systems. *Recent Chem. Prog.,* 10:100–114.
10. Frankenhaeuser, B. (1960): Sodium permeability in toad nerve and in squid nerve. *J. Physiol. (Lond.),* 152:159–166.
11. Gilbert, D. L., and Ehrenstein, G. (1969): Effect of divalent cations on potassium conductance of squid axons: Determination of surface charge. *Biophys. J.,* 9:447–463.
12. Ginsborg, S., and Noble, D. (1976): Use of current-voltage diagrams in locating peak energy barriers in cell membranes. *J. Membr. Biol.,* 29:211–229.
13. Goldman, D. E. (1943): Potential impedance and rectification in membranes. *J. Gen. Physiol.,* 27:37–60.
14. Hall, J. E., Mead, C. A., and Szabo, G. (1973): A barrier model for current flow in lipid bilayer membranes. *J. Membr. Biol.,* 11:75–97.
15 Heckmann, K., Lindemann, B., and Schnakenberg, J. (1972): Current-voltage curves of porous membranes in the presence of pore-blocking ions. I. Narrow pores containing no more than one moving ion. *Biophys. J.,* 12:683–702.
16. Hille, B. (1975): Ionic selectivity of Na and K channels of nerve membranes. In: *Membranes,* Vol. 3, edited by G. Eisenman, pp. 255–323. Dekker, New York.
17. Hille, B. (1975): Ion selectivity, saturation and block in sodium channels—a four barrier model. *J. Gen. Physiol.,* 66:535–560.
18. Hille, B., Woodhull, A. M., and Shapiro, B. I. (1975): Negative surface charge near sodium channels of nerve: Divalent ions, monovalent ions, and pH. *Philos. Trans. R. Soc. Lond. [Biol.],* 270:301–318.
19. Hodgkin, A. L., and Katz, B. (1949): The effect of sodium ions on the electrical activity of the giant axon of the squid. *J. Physiol. (Lond.),* 108:37–77.
20. Jan, L. Y., and Jan, Y. N. (1976): L-glutamate as an excitatory transmitter at the *Drosophila* larval neuromuscular junction. *J. Physiol. (Lond.),* 262:215–236.
21. Laüger, P. (1973): Ion transport through pores: A rate-theory analysis. *Biochim. Biophys. Acta,* 311:423–441.
22. Lewis, C. A. (1979): The ion concentration dependence of the reversal potential and the single channel conductance of ion channels at the frog neuromuscular junction. *J. Physiol. (Lond.),* 286: *(in press.)*
23. Moore, W. J. (1972): *Physical Chemistry, 4th ed.,* p. 372. Prentice-Hall, Englewood Cliffs, N.J.
24. Muller, R. U., and Finkelstein, A. (1974): The electrostatic basis of Mg^{++} inhibition of transmitter release. *Proc. Natl. Acad. Sci. USA,* 71:923–926.
25. Neher, E., and Sakmann, B. (1976): Noise analysis of drug induced voltage clamp currents in denervated frog muscle fibres. *J. Physiol. (Lond.),* 258:705–729.
26. Noble, D. (1972): Conductance mechanisms in excitable cells. In: *Biomembranes,* 3, edited by F. Kreuzer and J. F. G. Slegers, pp. 427–447. Plenum Press, New York.
27. Robinson, R. A., and Stokes, R. H. (1960): *Electrolyte Solutions.* Butterworths, London.
28. Sandbloom, J. P., and Eisenman, G. (1967): Membrane potentials at zero current: The significance of a constant ionic permeability ratio. *Biophys. J.,* 7:217–242.
29. Sandbloom, J., Eisenman, G., and Neher, E. (1977): Ionic selectivity, saturation and block in Gramicidin A channels: I. Theory for the electrical properties of ion selective channels having two pairs of binding sites and multiple conductance states. *J. Membr. Biol.,* 31:383–417.
30. Schauf, C. L. (1975): The interaction of calcium with *Myxicola* giant axons and a description in terms of a simple surface charge model. *J. Physiol. (Lond.),* 248:613–624.
31. Shatkay, A. (1968): Individual activity of calcium ions in pure solutions of $CaCl_2$ and in mixtures. *Biophys. J.,* 8:912–919.
32. Takeuchi, N. (1963): Some properties of conductance changes at the end-plate membrane during the action of acetylcholine. *J. Physiol. (Lond.),* 167:128–140.
33. Takeuchi, N. (1963): Effects of calcium on the conductance change of the end-plate membrane during the action of transmitter. *J. Physiol. (Lond.),* 167:141–155.
34. Takeuchi, A., and Takeuchi, N. (1960): On the permeability of end-plate membrane during the action of transmitter. *J. Physiol. (Lond.),* 154:52–67.

35. Teorell, T. (1953): Tansport processes and electrical phenomena in ionic membranes. *Progr. Biophys. Chem.,* 3:305–369.
36. Woodbury, J. W. (1971): Eyring rate theory model of the current-voltage relationships of ion channels in excitable membranes. In: *Chemical Dynamics* edited by J. O. Hirschfelder and D. Henderson, pp. 601–61F. Wiley (Interscience), New York.
37. Woodhull, A. M. (1973): Ionic blockage of sodium channels in nerve. *J. Gen. Physiol.,* 61:687–708.

Subject Index